Towards a Renewable Hydrogen Strategy for Mongolia

OECD
BETTER POLICIES FOR BETTER LIVES

Please cite this publication as:
OECD (2025), *Towards a Renewable Hydrogen Strategy for Mongolia*, OECD Publishing, Paris, https://doi.org/10.1787/15122489-en.

ISBN 978-92-64-75137-8 (print)
ISBN 978-92-64-54230-3 (PDF)
ISBN 978-92-64-37566-6 (HTML)

Photo credits: Cover © witsarut sakorn/Shutterstock.com.

Corrigenda to OECD publications may be found at: https://www.oecd.org/en/publications/support/corrigenda.html.

Foreword

This report follows from a request from the Ministry of Economy and Development of Mongolia to the OECD Sustainable Infrastructure Programme in Asia (SIPA) for support in the elaboration of a national strategy for the development of a renewable hydrogen industry. SIPA is financed by the German Federal Ministry for the Environment, Nature Conservation, Nuclear Safety and Consumer Protection and the International Climate Initiative and is implemented by a consortium of partners led by the OECD.

This report was scoped at a public-private roundtable that was held in Ulaanbaatar in March 2023 and co-organised between the OECD Sustainable Infrastructure Programme in Asia (SIPA) and the Mongolian Hydrogen Council, the Ministry of Energy, and the Ministry of Economy and Development. The preliminary findings were discussed with the Ministry of Economy and Development and Ministry of Energy during a mission to Ulaanbaatar in October 2023, which was followed with a public-private workshop in April 2024 at which the overall assessment and recommendations of the report were presented and validated. The present report has been revised following the April 2024 workshop to reflect recent policy and project developments, as well as the opinions and observations of local stakeholders.

The findings of the report are relevant to policymakers and firms in Mongolia that wish to develop and take advantage of new opportunities in the context of renewable hydrogen and its value chains. At the same time, the report and its recommendations are also relevant for all stakeholders interested in the low-carbon transition of the Mongolian economy. As discussed throughout this report, this is because the establishment of the general framework conditions necessary for a less carbon-intensive and more environmentally sustainable development of industry are a pre-condition for the adoption of more advanced, frontier technology options such as hydrogen specifically.

Acknowledgements

This report was prepared under the OECD Sustainable Infrastructure Programme in Asia (SIPA) and managed by Peline Atamer, Head of SIPA in Central Asia. The report benefited from the strategic guidance of William Tompson, Head of the Eurasia Division of the OECD Global Relations Secretariat, and Yuval Laster, Head of the Finance, Investment and Global Relations Division of the OECD Environment Directorate. The report was developed in coordination with Balchinluvsan Lkhagvasuren, Director General of the Integrated Policy and Planning Department at the Ministry of Economy and Development.

The authors of the report were Luke Mackle and Peline Atamer of the OECD SIPA. The report benefited from research and analytical support from Francesca Al Toum and Maggie Keyworth, Junior Analysts with OECD SIPA, and from Prof. Basandorj Davaa, expert, as well as Marijn Korndewal and Lylah Davis of the OECD Environment Directorate for Chapter 4. The report would not have been possible without the exceptional support of Enkhbayar Dambadarjaa, Local Consultant to SIPA in Ulaanbaatar, and Elisa Larrakoetxea, Project Administrator with SIPA in Paris.

The SIPA team are grateful to the various stakeholders in Mongolia and colleagues at the OECD and in other organisations who participated in consultations for the preparation of this work and provided insightful comments and guidance on the report as it developed, notably: Krzysztof Michalak, Virginie Marchal, Douglas Herrick, Deger Saygin, Joseph Cordonnier, Moongyung Lee, Matthew Griffiths and Takayoshi Kato of the OECD Environment Directorate; Byambasaikhan Bayanjargal and Itgel Yondongombo of the Mongolian Hydrogen Council; Matilda Dimovska, UNDP Resident Representative in Mongolia, Lin Cao, UNDP Deputy Representative, and Delgernaran Tumurtogoo, UNDP National Project Coordinator; Makhbal Tumenjargal, Director of the Mongolian Energy Economics Institute; Saintulga Munkhgerel of Decarbonisation Manager at Oyu Tolgoi; Prof. Ulemj Damiran of the Mongolian University of Science and Technology; Anna Nilsson of the New Climate Institute; Mikaa Blugeon-Mered of Sciences Po; Dr. Dunja Hofmann and Lukas Dietrich of GIZ; Baldorj Myagmarsuren, Ministry of Energy of Mongolia; Dagva Gankhuyag and Khishigt Tamir of Newcom Group; Dr. Sevjidsuren Galsan of the Mongolian Academy of Sciences; Byambasaikhan Bayanjargal and Neil Young of the Gobi H2 project; Dr. Oliver Schnorr, David Pauli, Gunchin-Ish Khosbayar, and Khangalsaikhan Bayaraa of EuroKhan; Zula Luvsandorj, Galtbaatar Davaasuren, and Mike Buck of the SunSteppe H2 pilot project; Zoljargal Balsandorj, Tuguldur Mungunselem and Temuulen Purevdorj of Mongolyn Alt (MAK).

Table of contents

FIGURES

TABLES

Executive summary

The global push for net zero could create economic opportunities for Mongolia as a producer of renewable hydrogen

As a signatory to the Paris Agreement on Climate, Mongolia has joined the international effort to reach net zero emissions by the year 2050. Achieving this ambitious goal will require a massive deployment of existing and new technologies across a wide array of industrial activities to reduce the carbon intensity of economic output. One such technology is hydrogen, and in particular renewable hydrogen produced using electrolysis powered by renewable energy sources. Renewable hydrogen has the particular advantage of being a viable alternative to fossil fuels in some of the hardest-to-abate industrial activities, such as metallurgy, cement, chemicals production, and mining. If forecasts from the International Energy Agency (IEA) are correct, then global hydrogen production will need to rise from 94 megatons (Mt) of hydrogen in 2021 to 500 Mt of hydrogen by 2050, of which 90% must come from renewable or low-carbon sources. Almost all of the capacity necessary to produce this renewable hydrogen is yet to be installed.

As the international market for renewable hydrogen matures and demand for the technology increases, there may arise significant opportunities for emerging economies such as Mongolia. Most international organisations and industrial stakeholders assume that for renewable hydrogen to be an economically competitive alternative to fossil fuels then it must be produced for less than USD 3 per kilogram. The drivers of that cost are varied, with some related to location (and by extension, distance to market), and others determined by inputs. The two most significant cost components are electrolysers and the levelised cost of renewable electricity. As the technology matures and is more widely deployed in the coming years, the cost of electrolysers is widely expected to come down, leaving the cost of renewable electricity as the major cost determinant. Given Mongolia's exceptionally high renewable energy potential, it is here that the country may be able to distinguish itself as a competitive renewable hydrogen producer.

The potential for Mongolia to be a producer of renewable hydrogen has already attracted private sector interest. Drawn by the quality of Mongolia's potential for combined solar and wind power generation, which is overwhelmingly located in the South Gobi region of the country, investors – both new to Mongolia and established – are at various stages of developing renewable hydrogen projects. Two sets of considerations are driving these first movers to explore the development of a frontier low-carbon technology in Mongolia. The first is the aforementioned quality of the renewable energy potential, the second is the relative proximity to a large potential export market, namely the steel industry in northern China. As discussed in this report, while the domestic market for renewable hydrogen remains at a very early stage of development, developers have already begun to secure off-take contracts with major industrial enterprises within Mongolia, with the decision making of these firms in large part driven by external pressure and encouragement to decarbonise their value chains, regardless the maturity of the low-carbon transition in the countries in which their operations are located.

A competitive renewable hydrogen industry in Mongolia could support diversification and decarbonisation

There are multiple rationales for Mongolia to explore the production and use of renewable hydrogen. If Mongolia is able to successfully develop its vast renewable energy potential, then it can feasibly produce renewable hydrogen at a sufficiently low cost to be a competitive exporter. Similarly, given the nascent stage of development of the many of the likely future applications of hydrogen – from fertilisers to synthetic fuels – there is an opportunity for Mongolia to position itself if new low-carbon value chains and thereby support the long-standing commitment to export diversification.

The implications for Mongolia producing low-cost renewable hydrogen might be greatest for the country's mining sector, where both what is mined and how it is mined is changing. The government has well-established plans to develop the country's industrial sector in general and its mining sector in particular, with ambitions to cement its position as a locum of investment. Developments in the mining sector are a bellwether of the challenges and opportunities that the global push for decarbonisation will present to Mongolia. On the one hand, with the decision at COP 28 in 2023 to transition away from fossil fuels, the future of Mongolia's coal exports is uncertain. On the other, the push for electrification continues to raise the importance of the country's important copper reserves, as well as its undeveloped reserves of critical rare minerals.

Mongolia has an opportunity to ensure that mining remains an economically important activity in the context of the green transition. Nevertheless, without significant a significant policy push to embed low-carbon technologies such as hydrogen in infrastructure investment decisions in the mining sector, there is a risk that Mongolia's industrial development will be incompatible with the government's broader sustainability ambitions. The mining sector, a significant driver of rising demand for power in Mongolia's overwhelmingly coal-fired electricity generation system, can be a driver of decarbonisation elsewhere in the long-term only if it begins to address its sustainability domestically. Given the global push to decarbonise entire mining value chains, if Mongolia moves into higher value-adding activities such as smelting, then the need for alternatives to coal for power will only grow. Renewable hydrogen is one technology that could play an important role in this process.

Towards a renewable hydrogen strategy for Mongolia: Building on private sector interest, aligning with Mongolia's low-carbon objectives

Recognising both the opportunities and challenges that the development of a renewable hydrogen industry presents, the Government of Mongolia has taken some initial stages towards formulating a hydrogen strategy. As discussed throughout this report, should Mongolia move forward with developing a renewable hydrogen industry, there will likely be a clear need for government leadership. The Ministry of Economy and Development was the first institution to explore the possibility of a national strategy for hydrogen development, responding in part to the private sector interest that has been consistently articulated through the Mongolian Hydrogen Council. Other government institutions, the Ministry of Energy and the Industrialisation Policy Committee among them, have since began to explore the development of hydrogen production in Mongolia, as well as its use in existing and new industrial applications.

A priority for policymakers will be balancing the ambition to develop a national hydrogen industry with ensuring that policy support does not crowd out investment and innovation in other low-carbon areas. While Mongolia has natural resources that could be leveraged to become an important producer of renewable hydrogen, the development of that market remains uncertain, and domestically a number of the hypothetical off-takers in the mining sector are pursuing other technologies in parallel for their decarbonisation. The challenge is therefore to simultaneously improve the effectiveness of innovation and industrial policy support for renewable hydrogen in a way that does not detract from private sector initiatives in other low-carbon technology domains. To this end, 'technologically neutral' innovation and industrial policies are likely to play an important role in any policy toolkit for the development of renewable hydrogen capacities in Mongolia, as well as "no regret" policies to improve the broader enabling environment for decarbonisation in the country, such as the development of Mongolia's vast renewable energy resources for power sector decarbonisation. There is also a balance to be struck between committing resources that may improve Mongolia's competitiveness as a first mover in global renewable hydrogen market (e.g., providing significant fiscal incentives to project developers) with taking a more cautious approach and committing public resources for hydrogen development as the global market matures and technology costs decrease.

Any low-carbon economic transition presents systemic challenges, and one of the greatest obstacles for governments is the difficulty in aligning the interests of the manifold economic actors and policy domains whose buy-in and input are necessary for success. Given the early stage of Mongolia's low-carbon transition, the emergence of private sector interest to develop renewable hydrogen production in the country, despite the many challenges this implies, may come at a propitious moment. Mongolia's renewable hydrogen ambitions can act as a locum for the broader process of industrial decarbonisation that is necessary to ensure the long-term sustainability and competitiveness of the economy. Should Mongolia be successful in developing this nascent industry, then the decisions taken over the coming years on infrastructure, technology and capacities will shape the extent to which the domestic industrial sector benefits from its development, as well as the environmental sustainability of this development. However, should Mongolia's hydrogen ambitions be jettisoned, circumscribed or postponed, then the actions outlined in this report, based on the analysis presented throughout, will nevertheless support the government's broader low-carbon transition objectives, since they will accelerate the rollout of the knowledge, skills, technologies and incentives necessary to set Mongolia's growing industrial sector on a more sustainable pathway.

Four areas for action and strategic coordination to ensure a renewable hydrogen industry works for Mongolia

Attracting investment to develop a renewable hydrogen industry

A renewable hydrogen industry at scale in Mongolia will require a massive mobilisation of external investment in technology and infrastructure, while FDI has historically been predominantly concentrated in the mining sector. There is a need for the government to make a case for why Mongolia, and why now. A clear example of this is in renewable energy infrastructure, both for production and transmission. In the case of the former, just one of the projects being explored in Mongolia envisages the installation of 600 megawatts (MW) of renewable power, more than double the renewable capacity currently installed in the country. Yet, investment into the country's renewable energy infrastructure, for all its obvious potential, has not been forthcoming; partly this is a question of incentives, partly it is one of risk. Investors perceive risks in the country, particularly in the types of investor-state co-operation that is emerging as crucial to getting early hydrogen projects off the ground, such as off-take and hydrogen purchasing agreements. There is therefore a clear need to raise investor confidence. There is also a need to raise the profile of Mongolia as a potential location for renewable hydrogen production and ensuring that the country's investment climate more broadly does not dampen investors' appetite to pick Mongolia over the number of other emerging economies where similar projects are being developed.

Aligning innovation and technology adoption with low carbon outcomes

While there is scope for hydrogen to play a major role in the decarbonisation of Mongolian industry, the preparedness of Mongolian industry to use low-carbon technologies remains very low. At the same time, the market signals that constitute a major part of the policy framework for creating domestic demand for the development and deployment of such technologies – effective carbon pricing, conditional public procurement, directed public research and development (R&D) support, quotas – have yet to be established. Where innovation has taken place in low-carbon technologies, it has generally happened within large multinational mining companies which face external pressure to decarbonise and increase the sustainability of their value chains, with these companies benefiting from their own well-developed international networks of R&D. In other instances, hydrogen-related innovation is taking place, both in the private sector and in state-owned enterprises (SOEs) but is linked to government ambitions to add greater value to the country's significant coal extraction industry; such initiatives are technologically advanced and a testament to the significant innovation capacities in Mongolia's leading industries, but the output of their innovation expenditure risks being misaligned with the emerging standards and expectations of global markets for renewable hydrogen and its derivatives.

Ensuring that water use for a renewable hydrogen industry is sustainable

In addition to renewable electricity, the other critical input necessary for renewable hydrogen production is water, and water of a sufficiently high purity. Mongolia is a country facing water scarcity: not necessarily at the country level, but at the regional level, and this has significant implications for the development of a renewable hydrogen production industry at scale. There is a significant geospatial mismatch between the location of country's freshwater resources, which are primarily in the north, and where the most propitious renewable electricity generation potential is found, which is in the south of the country. In the South Gobi, where all renewable hydrogen pilot projects are currently located, and where the water-intensive mining sector is also concentrated, the only source of water are saline aquifers, the reserves and replenishability of which remain uncertain. Where other emerging economies can explore the possibility of seawater desalination for hydrogen production, Mongolia evidently cannot. The development of the country's renewable electricity potential in the south will be crucial to meeting the government's broader low-carbon objectives, but the extent to which they can be utilised for hydrogen production will depend on whether projects can identify sustainable sources of water, developing the necessary infrastructure to utilise these sources, and ensuring that this use is environmentally, economically and socially sustainable.

Building a national vision and strategy for hydrogen development will require to establish a platform that brings together stakeholders from across the economy

As in most countries envisaging the production and use of renewable (or sometimes more broadly low-carbon) hydrogen, developing a national vision and strategy for such development will require putting in place proper steering and coordination mechanisms. Mongolia can build on the public-private working group that the OECD and UNDP have established and consulted for this work, which brings together several ministries, private sector representatives including the Mongolian Hydrogen Council, national experts and academics, and international partners. Working groups can be established to drive the agenda in specific areas such as innovation and technology transfer, water and hydrogen, infrastructure, and standards and regulations. The establishing of such a platform for a national industrial policy dialogue about renewable hydrogen is important in many respects: it enables to bring together knowledge and information that is disseminated across stakeholders, it helps identify critical problems and identify solutions, including and importantly those implying policy and regulations. It also enables to increase institutional knowledge about a new green technology, its implications, and its interaction with the broader policy framework for economic growth and decarbonisation in Mongolia.

1 Introduction: A renewable hydrogen strategy in support of Mongolia's low-carbon transition

This chapter introduces the rationale for developing a renewable hydrogen strategy in Mongolia. The country's economic development has been driven mainly by the mining of coal and other minerals, activities with high environmental and social costs. The importance of mining and the significant role of coal in its power sector will likely make it difficult for the country to meet its emissions reduction commitments under the Paris Agreement. As Mongolia explores the potential of greener technologies and private sector renewable hydrogen projects emerge in the South Gobi Desert, the government has an opportunity to establish a national renewable hydrogen strategy. This strategy would support these early projects and lay the foundations for low-carbon technology development. By steering the production and adoption of renewable hydrogen, it could help decarbonisation Mongolia's economy and facilitate its transition to a more sustainable and resilient development pathway.

Mongolia's economic model poses challenges to its low-carbon transition

Mongolia's ability to deliver on its low-carbon commitments, and more broadly, on the transformation needed to put the country on a more resilient, and environmentally sustainable economic pathway, will require it to address two intertwined challenges. For the foreseeable future, Mongolia's economic growth will be driven by the export of coal and other mineral products that the country has in abundance, which is a highly emission-intensive activity. Secondly, the country's power system relies heavily on coal, and Mongolia is one of the most energy-intensive countries in the world. In spite of its commitments to the Paris Agreement on Climate Change (Paris Agreement) and associated Nationally Determined Contribution (NDC) targets, along with the adoption of plans and policies in recent years aimed at decarbonising the economy, the clean energy transition in Mongolia has not yet begun. This section looks in detail at these challenges to Mongolia's decarbonisation.

Over the past 20 years Mongolia's economic growth has been driven primarily by the export of minerals

Since exiting a deep recession in the early 1990s, Mongolia's economic growth has been consistently robust, averaging 5.9% between 2000 and 2022 (IMF, 2023[1]). Between 1998 and 2018, the share of the population living in poverty dropped from 28.5% to 0.7% (USD 2.15 per day threshold) (World Bank, 2023[2]), and GDP per capita (PPP) has risen from 13.1% of the OECD average in 2000 to 28.3% in 2022; however, the rate of convergence has slowed in recent years (World Bank, 2023[3]). Since 2000, productivity across all sectors of the economy has also grown significantly, rising from USD 1,875 per worker in agriculture in 2000 to USD 5,705 per worker in 2019, from USD 9,395 to USD 14,156 per worker in industry, and USD 6,673 per worker (2010) to USD 9,516 per worker in services, with productivity growth in this latter category more muted than elsewhere (World Bank, 2023[3]).

This growth has been fuelled primarily by the extraction and export of a narrow range of mineral products that the country has in abundance – principally copper, gold and coal. Some 70.1% of Mongolia's exports consist of unprocessed mineral products. One market – China – accounts for 73% of Mongolia's exports, with Mongolian exports to that country growing 131% over 2011-2021 (OEC, 2023[4]). In 2021, 38% of Mongolian exports to China consisted of copper ore; coal's share was 32.1% coal, followed by iron ore (12.4%), crude petroleum (3.58%) and zinc ore (2.23%) (OEC, 2023[4]). Mongolia has the 24th largest total proven reserves (2.5 billion tonnes) in the world, and it is the 14th largest coal producer in the world (892,455 Terajoule (TJ)of coal extracted in 2021, down from a peak of 1,123,854 TJ in 2019). At the national level, the economic and energy significance is clear. In 2022, Mongolia produced the equivalent of 61,257 kilowatt-hours (kWh) of coal per capita, second to only Australia (121,618 kWh). According to a 2020 IMF study on the economic implications of coal dependence for coal producers in the energy transition, coal production was equivalent to 30% of the country's GDP. Of this production, the Fund estimated the net export of coal was equivalent to 22.5% of GDP. In both cases, Mongolia stands significantly ahead of the other surveyed countries (BP, 2022[5]) (IMF, 2022[6]).

Given the absence of a large domestic market, growth is hugely dependent on external demand, increasingly from China. The vast majority of Mongolia's domestic industrial and manufactured output is exported, with exports of goods and services in 2022 equal to 69.7% of GDP. This percentage is significantly higher than the OECD average of 30.4% and exceeds that of other countries in the region with large extractive sectors, such as Kazakhstan, where the equivalent figure was 33.5% (World Bank, 2023[3]). The situation in Mongolia is similar to that in other transition economies in the Eurasia region, where the small size of the domestic market means that external trade plays a major role in domestic output.

This extreme concentration of export products and trade partners increases vulnerability to external shocks. When the COVID-19 pandemic prompted an international shutdown of economic activity, Mongolia was one of the worst-affected economies in Central Asia, with GDP contracting 4.6% in

2020, second only to Kyrgyzstan, which saw its economy shrink 8.4% (World Bank, 2023[3]). The main driver of the shock to Mongolia's economy was the immediate passthrough effect of the decision by China to close its borders, which caused major disruptions to industrial output and trade (World Bank, 2023[7]). In 2020, slowdown in the mining sector was the largest single contributor to the country's economic contraction, accounting for 31.3% of the drop in GDP (National Statistical Committee of Mongolia, 2022[8]).

The mining sector is and will remain important to the country's economic development under current conditions. Mining output has picked up since China re-opened its borders in 2022, with the sector's output accounting for 25.8% value added that year, a far larger share than in other SIPA project countries in Central Asia, such as Kazakhstan (15.6%) and Uzbekistan (4.3%) (Figure 1.1). Between 2011 and 2022, mining accounted for 72.8% of all foreign direct investment inflows to Mongolia (FDI) (National Statistics Office of Mongolia, 2023[9]). The contribution of the mining sector to Mongolia's economy is expected to grow in the coming years as underground extraction begins at the country's vast Oyu Tolgoi (OT) copper and gold mine, with production forecast to double between 2023 and 2025 (World Bank, 2023[10]). Trade partners' interest in accessing minerals within Mongolia's subsoil, including critical minerals like copper and rare earths, as well as other minerals like uranium, could spur additional mining expansion. In 2011, Mongolia's minerals wealth was estimated between USD 1 and 3 trillion (Brian S. Fisher, 2011[11]). National plans to develop the sector are discussed in Chapter 3 of this report.

Figure 1.1. The structure of output in Mongolia, regional peers and the OECD area

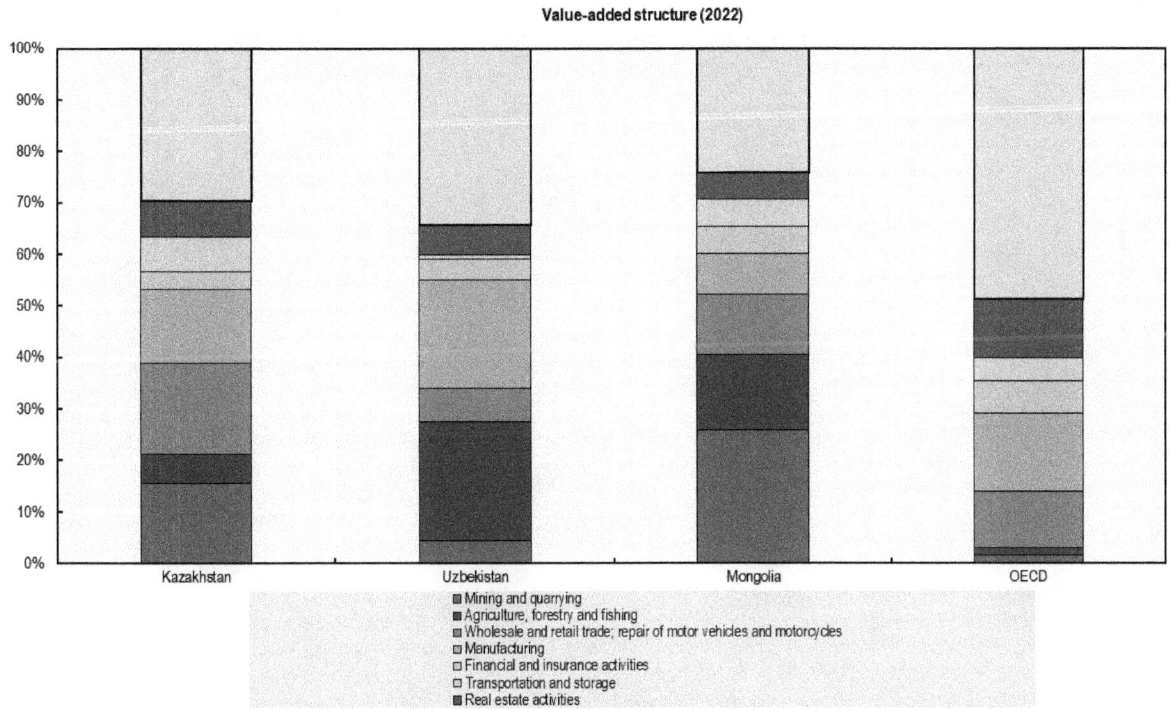

Note: Economic activities according to ISIC classifications. 'Other' contains the following categories: Public administration and defence; compulsory social security; Education; Construction; Human health and social work activities; Information and communication; Electricity, gas, steam and air conditioning supply; Professional, scientific and technical activities; Accommodation and food service activities; Administrative and support service activities; Water supply; sewerage, waste management and remediation activities; Other service activities; Arts, entertainment and recreation; Activities of households as employers; undifferentiated goods- and services-producing activities of households for own use

Source: National Statistical Agencies of Mongolia, Uzbekistan, and Kazakhstan. (OECD, 2021[12])

Although the industrial sector does not have as large a direct GHG footprint as the power sector (Figure 1.2a and Box 1.1), **it is a major source of GHG emissions and other forms of pollution** (IEA, 2022[13]). Industry accounts for 6% of Mongolia's total GHG emissions and the majority originates in the mining sector, with 95% of industrial emissions attributable to mining activities (Figure 1.2); this is significantly lower than the OECD average of 10.5% but higher than in Kazakhstan (3.1%) and similar to Uzbekistan (6.1%) (UNFCCC data put this at a slightly lower level, at around 3% of total emissions for Mongolia, but the trends are broadly consistent) (Gutschow et al., 2022[14]). As output from the sector has grown over the past 20 years, direct emissions have risen precipitously. The mining sector in Mongolia is also characterised by a high degree of carbon intensity, in large part due to the extremely high share of fossil fuels used to power mining operations. Mining in Mongolia produces USD 0.15 of value added (VA) per kilogram (KG) of CO2 emitted, compared an OECD average of USD 0.36 per KG of CO2 (Fig 1.c). When examining the relationship between mining growth and electricity demand, the impact of mining on Mongolia's emissions profile becomes notably pronounced. The country is currently constructing additional coal-fired generation to satisfy rising electricity demand from mining activities in the south of the country.

A similar story is true of the industrial sector's energy intensity more generally, with mining in Mongolia producing USD 2.2 of VA per megajoule (MJ) of energy consumed compared to an OECD average of USD 6.35 per MJ of energy (.Figure 1.1d). The difference with major mining economies such as Australia and Canada is even starker, where the equivalent figures are USD 0.52 VA per KG of CO2/USD 6.01 per MJ and USD 1.52 VA per KG of CO2/USD 27.3 per MJ. The sector is therefore not only very energy intensive, but also inefficient. There is likely considerable room for improvement in enhancing the energy efficiency of Mongolian economic performance. Upgrading industrial technologies and processes could yield significant improvements, especially at the lower end of the marginal abatement cost curve

Figure 1.2. Mining emissions in Mongolia and the carbon intensity of industrial output

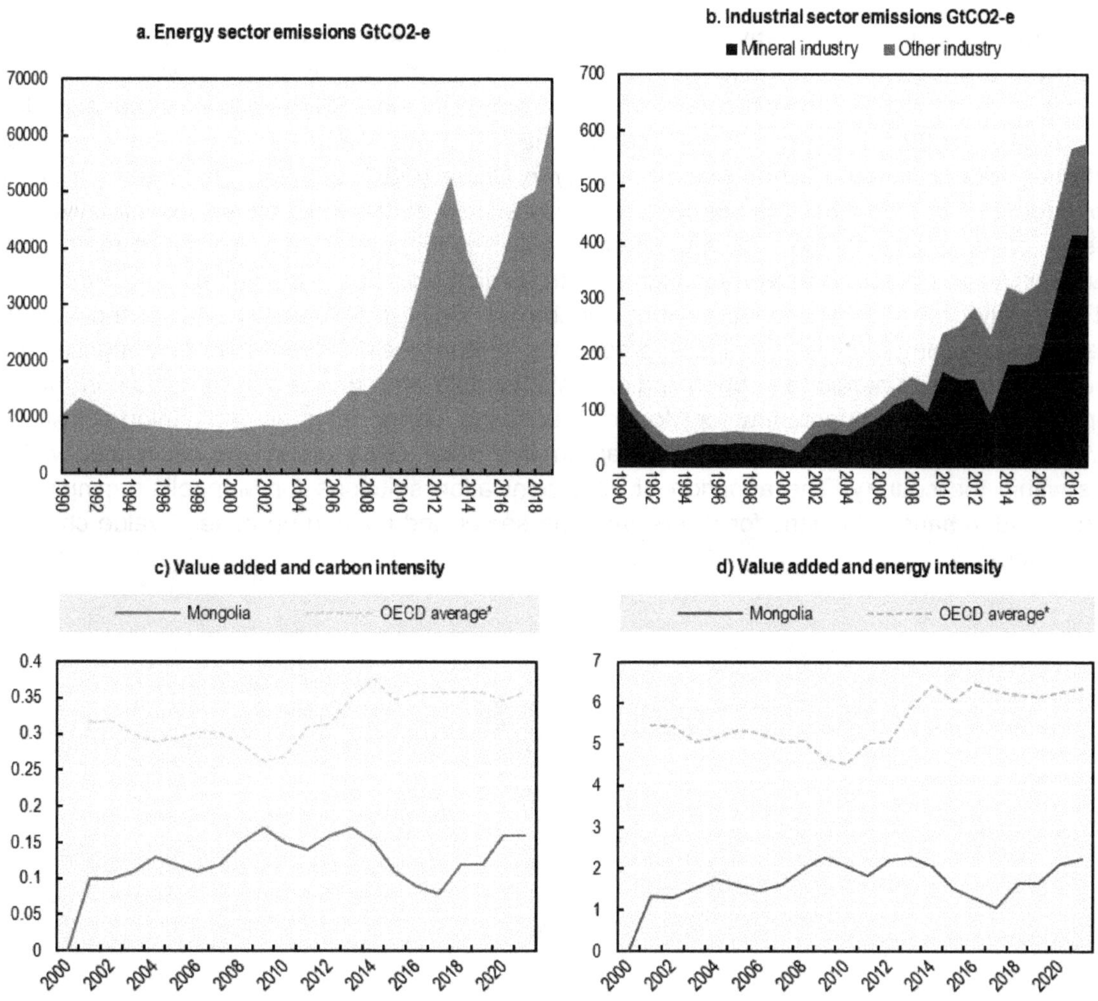

Note: In Figure 1.2c, the chart depicts the amount of value added in the mining sector per KG of CO2. In Figure 1.2d, the chart depicts the amount of value added in the mining sector per MJ of energy consumed.
Source: Figure 1.2a and b based on (Gutschow et al., 2022[14]); Figure 1.2c and d based on (IEA, 2023[15]).

Like a number of other areas of heavy industry, emissions from mining activities are considered to be "hard-to-abate". Such characterisations generally reflect the limited scope for emissions reductions through electrification, efficiency improvements, and readily available and commercially viable technological solutions. This is true in Mongolia as it is in other countries, but the significant gap between Mongolia and other mining economies in terms of the efficiency of energy use and carbon emissions relative to output highlights both the scope that the country has to improve the sustainability of mining activities as well as the ways in which other carbon intense activities – principally in the energy sector – have a pass-through effect on other sectors of the economy. The government's intention to significantly expand the scope of mining activities as part of its broader industrialisation and economic development strategy will have an impact on the emissions footprint of industry in Mongolia (Government of Mongolia, 2020[16]) (Government of Mongolia, 2023[17]). There is, therefore, a rationale to integrate sustainable and low-carbon technologies and approaches into industrial planning now – for example, supporting the embedding of low-carbon technology options into the development of Mongolia's nascent steel industry, exemplified by ongoing discussions in Mongolia to develop the Darkhan steel plant – which could help with the longer-term competitiveness of Mongolia's industry in a less carbon intensive international context.

Fostering the development of activities at higher stages in mining value chains will also have an impact on the domestic emissions intensity of the mining industry. At present, Mongolia exports the vast majority of its minerals in unprocessed form. For example, of the USD 2.9 billion copper exported from China in 2021 – an 80.5% increase over the previous five years – all of it was in the form of ores or concentrates, with processing taking place primarily in China (OEC, 2023[4]). This means that the true emissions footprint of the industry is not captured domestically as the most emissions-intensive activities are off-shored (as is currently the case with smelting activities for copper). It is the government's intention to expand mining activities and to develop higher value-adding activities in the value chain within Mongolia as part of its diversification and development agenda (cf. Point 1 of Objective 4.2 of the Vision 2050 document on increasing the level of mineral processing in Mongolia) (Government of Mongolia, 2020[16]). The needs for critical minerals to support a global sustainable and low-carbon transition, in this regard, may represent a sizeable opportunity for Mongolia. Locating higher value-adding mining activities on its territory would, however, also introduce the carbon and other GHG emissions associated with these activities within the country. The adoption of decarbonisation solutions for Mongolia's mining industry should go hand in hand with plans for diversifying the sector and moving up mineral value chains in the coming decades.

Box 1.1. Profile of Mongolia's greenhouse gas emissions: Insights from Mongolia's National Inventory Report to the UNFCCC

In 2020, Mongolia's total greenhouse gas (GHG) emissions were equal to 43,081.62 gigagrams of CO2-equivalent (Gg CO2-e), when excluding emissions from Land Use, Land Use Change and Forestry (LULUCF). This represents an 82.17% increase in GHG emissions since 1990; net GHG emissions (i.e., total emissions minus removal of emissions) were equal to 12,909.10 Gg CO2-e in 2020, which is equal to a 340.02% increase from the 1990 level (Figure 1.3).

Figure 1.3. Mongolia's emissions profile based on the National Inventory Report to the UNFCCC

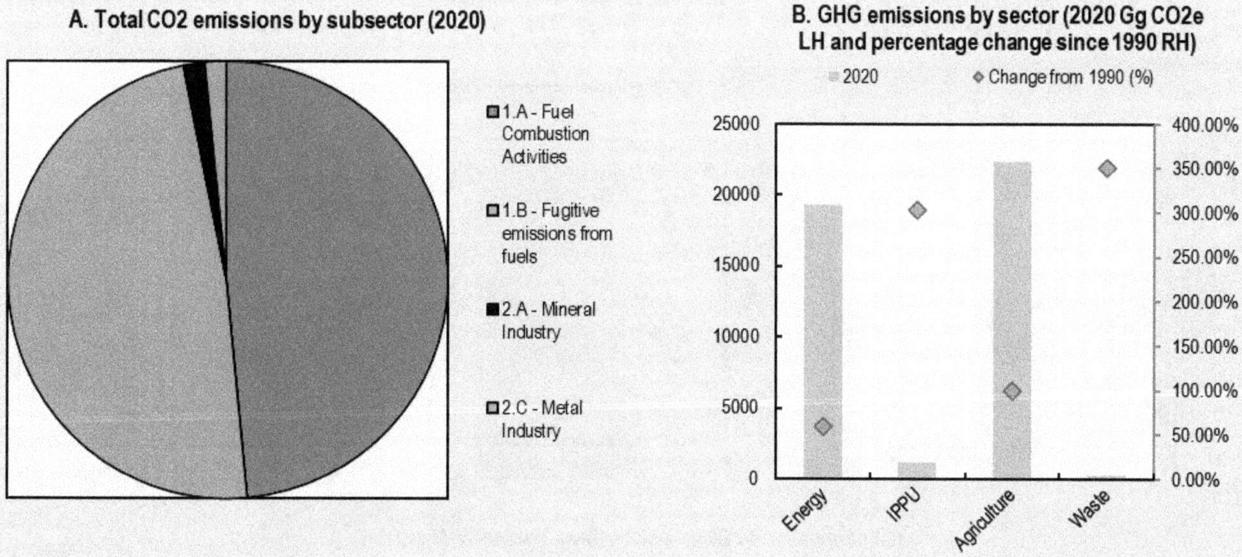

Note: For Panel A, the category Land Use, Land Use Change and Forestry (LULUCF) is excluded, the values being -30,172.52 GgCO2e in 2020 or a 3.95% change since 1990.
Source: (UNFCCC, 2023[18])

The agriculture and energy sectors are by far the biggest contributors to Mongolia's GHG emissions. In 2020, emissions from the agriculture sector accounted for 51.97% of total emissions, while the energy sector accounted for 44.78% of total emissions. Emissions from industrial processes were equivalent to 2.66% of total emissions, though this does not account for indirect emissions in industry through energy consumption; of industrial emissions, the mining and minerals sector accounts for 50.11% emissions, and total emissions from this sector continue to rise sharply.

Carbon dioxide (CO2) is the most important GHG in Mongolia, and represented 42.4% of total GHG emissions in 2020, followed by methane (33.82%) and nitrous oxide (22.46%). In 2020, Mongolia emitted 18,256.66 Gg CO2, a 52.59% increase from 1990. Of these emissions, 96.84% are accounted for by the combustion of fossil fuels for energy.

Source: (UNFCCC, 2023[18])

Mongolia's coal-powered energy system is a barrier to low-carbon, sustainable development

Mongolia has one of the most emissions-intensive economies in the world. Measured in terms of CO2 intensity of GDP (KG of CO2 per USD 1 PPP of output) or the energy productivity of GDP (GDP per unit of total energy supply), Mongolia significantly lags the OECD average (Figure 1.3a). In short, Mongolia requires more energy and emits more carbon to produce the same output as comparable OECD and non-OECD countries, with 1 tonne of oil equivalent (TOE) of energy in 2020 necessary to produce USD 6 102 output, compared to OECD average of USD 10 898 (Figure 1.3b) (OECD, 2023[19]).

Figure 1.4. CO2 and energy intensity of Mongolia's economy

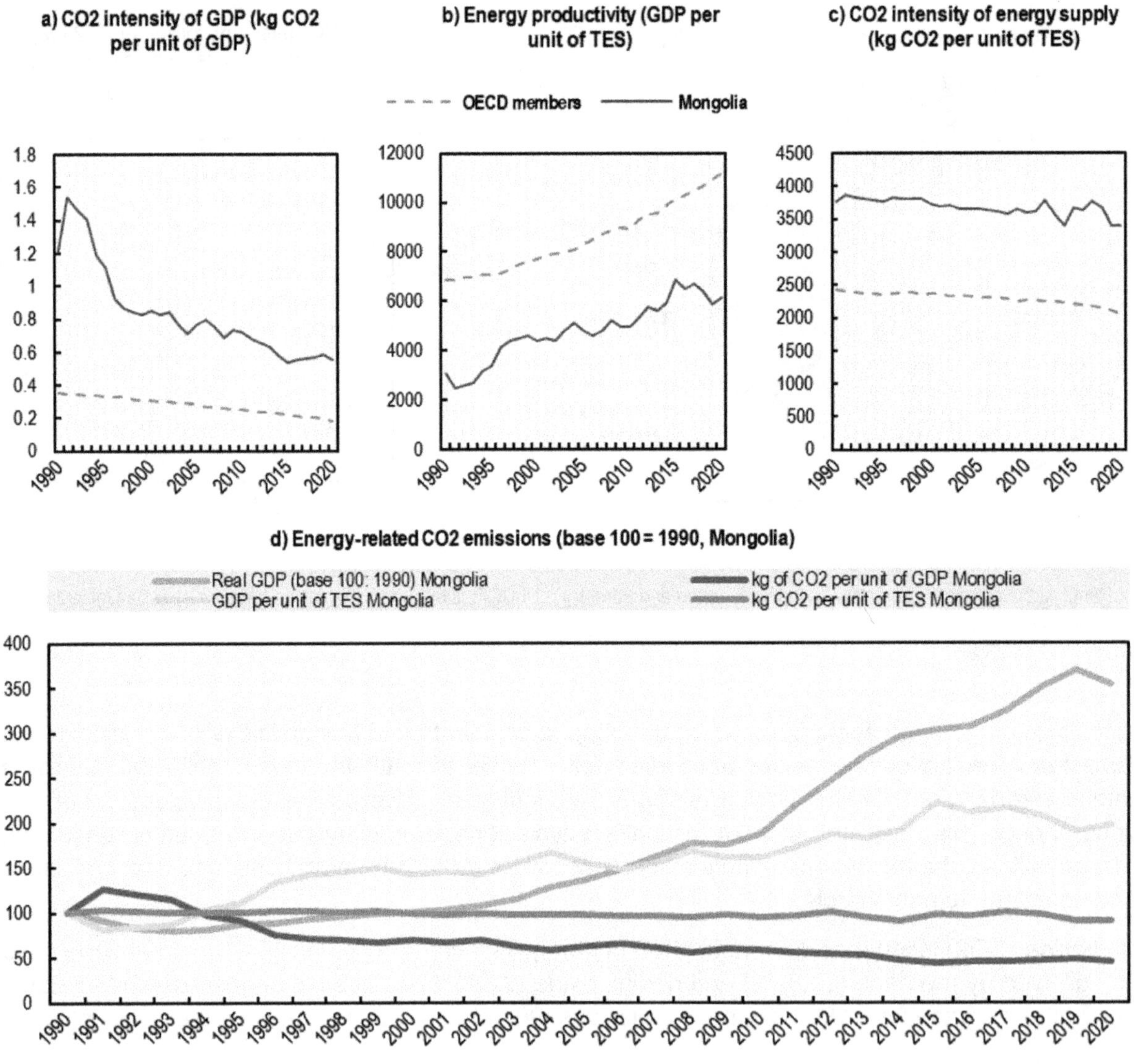

Note: TES = Total Energy Supply
Source : (OECD, 2023[20])

The persistently high levels of CO2 emissions and energy intensity of production is due to in large part to the dominance of fossil fuels in the country's energy supply. The power and heat sector accounts for 65% of Mongolia's CO2 emissions, with 85% of electricity generation being coal-fired (IEA,

2023[21]). There is a modest positive trend in most emissions intensity and growth indicators, including the CO2 intensity per unit of total energy supply (a key factor in the decarbonisation of the domestic energy sector), which indicates that Mongolia is gradually increasing the energy efficiency of the industrial base and energy sector. The CO2 intensity of energy supply nevertheless remains significantly higher than the OECD average, and further convergence of Mongolia with the OECD trend will require significant investment and reform in the country's energy sector.

In addition to environmental costs, there is a significant public health cost to the emissions-intensive nature of Mongolia's economy. The predominance of coal in Mongolia's energy supply is not only a major barrier to the country realising its own GHG reduction targets, but it is a major source of pollution and causes significant environmental and social externalities. In 2019, the OECD estimates that some 51% of Mongolia's population were exposed to dangerously high levels of particulate matter through air pollution compared to an OECD average of 0.45% (OECD, 2023[19]). The cost of Mongolia's pollution, in both public health and public finance terms, is significant. The OECD estimated that the welfare cost of premature deaths due to particulate matter in Mongolia was equal to 7.7% GDP in 2021, equivalent to 696 premature deaths per million of the population (OECD, 2023[22]).

Despite commitments in its Nationally Defined Contribution, Mongolia's energy transition has yet to begin

If not properly anticipated, the global transition to net zero can create challenges for Mongolia, particularly through the need to address the economic and social importance of coal. The Net Zero by 2050 scenario set out by the IEA envisages a precipitous decrease in coal demand over the coming decades, with demand for unabated coal use to fall by 98% by 2050 (IEA, 2021[23]). As global demand for coal falls, there will likely be an impact on coal-producing regions in terms of economic output and employment, as well as on public authorities through decreased tax revenues and export earnings on coal production; mining is found throughout the country, but many of the most economically important mines, including Tavan Tolgoi (coal) and Oyu Tolgoi (coal and copper) are located in Ömnögovi Aimag (region), which covers a vast part of the South Gobi. Ensuring that a pivot away from hydrocarbons such as coal does not have a harmful impact on social inclusion and welfare, particularly but not exclusively on extractive-based emerging economies, underpins significant academic and policy attention being paid to the need to secure a 'just' transition (OECD, 2022[24]) (IEA, 2023[25]).

In 2020, the government updated its Nationally Determined Contribution (NDC) targets to reduce GHG emissions from the energy sector by 49.4% by 2030 compared to its 2015 baseline. Mongolia's NDC commitments take place against a backdrop of numerous domestic policy commitments to increase the share of renewables in the country's energy system. The country's key energy-related policy is the 2015 *State Policy on Energy* (SPE), which set a goal of increasing renewable electricity capacity from 7.62% in 2014 to 20% by 2023, and then to 30% by 2030 as a share of total electricity generation capacity (IEA, 2023[21]). Prior to the recent NDC update, the government also set a goal of expanding hydropower capacity by 0.9GW capacity from a level close to zero, as well as increasing renewable capacity by 0.6GW from other sources (Government of Mongolia, 2019[26]) (Government of Mongolia, 2015[27]).

Despite these commitments, the majority of planned additional capacity is due to come from coal. The abundance of cheap and easily accessible coal has perpetuated a highly carbon intensive, polluting, and inflexible power and heating system, the emissions from which grow linearly with electricity demand. As industry continues to grow and with increasing residential electricity demand in the context of internal migration to Ulaanbaatar, demand on Mongolia's power system is rising every year. The vast majority of power expansion in Mongolia is envisaged to come from additional coal-fired plants. At present, the Ministry of Energy of Mongolia has plans to expand existing coal-fired plant capacity by 810 megawatts (MW), as well as creating an additional 750MW of new coal-fired capacity. Total installed coal capacity is anticipated to almost double in the second half of the 2020s, as a number of expansion projects at coal-fired CHPs and new CHPs – including the 450MW Tavantolgoi CHP – come online (Ministry of Energy of Mongolia, 2022[28]).

Moreover, grid and power system inflexibility are major challenges to the integration of renewable energy penetration and deployment of clean electricity. The power generation capacities of Mongolia's CHPs are generally inflexible and linked to demand for heat rather than electricity (Nilsson et al., 2021[29]). One consequence of power system and grid inflexibility is the difficulty of integrating electricity produced from variable renewable energy (VRE) sources, such as wind and solar, which acts an implicit barrier to electrification and its potential as a vector of decarbonisation. The limited flexibility of the grid is one of the major challenges to integrating even modest shares of VRE. These challenges are exacerbated by a lack of interconnectedness between Mongolia's power grid, which is both a reflection of geography and underinvestment (Nilsson et al., 2021[29]).

The position of the Mongolian authorities on the role of coal in the socio-economic future of the country is at odds with national emissions reduction commitments and will pose a fundamental challenge to the development of low-carbon industries such as renewable hydrogen. As discussed in Chapter 3, Mongolia retains a substantial pipeline of planned coal capacity additions, while the roll out of new renewable power generation capacity remains very slow. Trends in international climate negotiations, such as the decision taken at COP 28 to phase-down unabated coal power and phase-out fossil fuel subsidies, the agreement reached at the OECD Arrangement on Officially Supported Export Credits to cease providing support for unabated coal-fired power plants, as well as decisions by multilateral finance organisations such as the World Bank and the ADB to phase-out the financing of new coal-fired power generation all indicate that there is a need for Mongolia to begin planning for a post-coal future (UNFCCC, 2023[30]) (OECD, 2021[31]) (ADB, 2024[32]) (World Bank, 2023[33]). While revenues from coal exports may fall over the coming decades, the global low-carbon transition will create opportunities for Mongolia. Many of these opportunities will lie precisely in the new, more complex and at times higher value-adding industrial activities that the government wishes to develop. This includes mining metals such as copper and other rare earth minerals critical to the global transition, as well as in new low-carbon technologies such as hydrogen. As discussed throughout this report, what is important for the government to retain is that the development of these activities will likely depend on a clear signal to phase-out coal and to stimulate investment in low-carbon areas.

Renewable hydrogen in the global transition to Net Zero: Opportunities for Mongolia

The acceleration of national efforts towards a net zero global economy by 2050 has given additional impetus to Mongolia's long-standing environmental and social rationales for decarbonisation. Around the world, countries are designing and implementing policy programmes to support the decarbonisation of economic production, with the Paris Agreement– to which Mongolia is a signatory – setting a goal of limiting global temperature increase to 1.5C above pre-industrial levels by the end of the century. The challenge for all countries pursuing this goal is the inherent complexity in coordinating the low-carbon transition. Governments are increasingly expected to take an interventionist and steering role to facilitate systemic change domestically – often expressed through a renewed interest in industrial policy – and adapting to it internationally (Criscuolo et al., 2022[34]). The low-carbon transition is shaping existing markets and creating new ones which adds additional commercial and economic rationale for governments to act quickly in order to ensure competitiveness and wellbeing for future generations. To achieve its decarbonisation objectives, the Government of Mongolia is working with international partners, including the OECD, to develop a long-term vision for low-carbon economic development.

Renewable hydrogen will play a critical role in achieving net zero both globally, and across economies (Figure 1.4**ab**). While there are numerous approaches for renewable hydrogen production, one of the most promising is 'green' or renewable hydrogen, which is produced using electrolysis powered by renewable energy sources, and at scale it can contribute to global net zero targets in four important ways (Box 1.2). First, hydrogen is a promising alternative to fossil fuels in the energy sector – where its use is expected to grow markedly by 2050 – and industrial processes that require high temperatures (Figure 1.4c); the only by product of hydrogen combustion is water, and it could therefore contribute to improving the sustainability of hard-to-abate sectors where the technical viability of electrification is limited. Second, hydrogen is a necessary input for the development of fuel-cell based vehicles, which are likely to play a critical role in reducing emissions in logistics, the built environment, and in sectors such as mining. Third, hydrogen is a feedstock for a number of chemical industries, and therefore renewable hydrogen could significantly support the decarbonisation of many carbon-intensive products such as fertilisers whilst also enabling the emergence of new synthetic fuels. Lastly, as a storer of energy, hydrogen can help integrate energy produced from intermittent renewable electricity sources (Cammeraat, Dechezlepretre and Lalanne, 2022[35])(IEA, 2022b).

Figure 1.5. Hydrogen in Net Zero scenarios

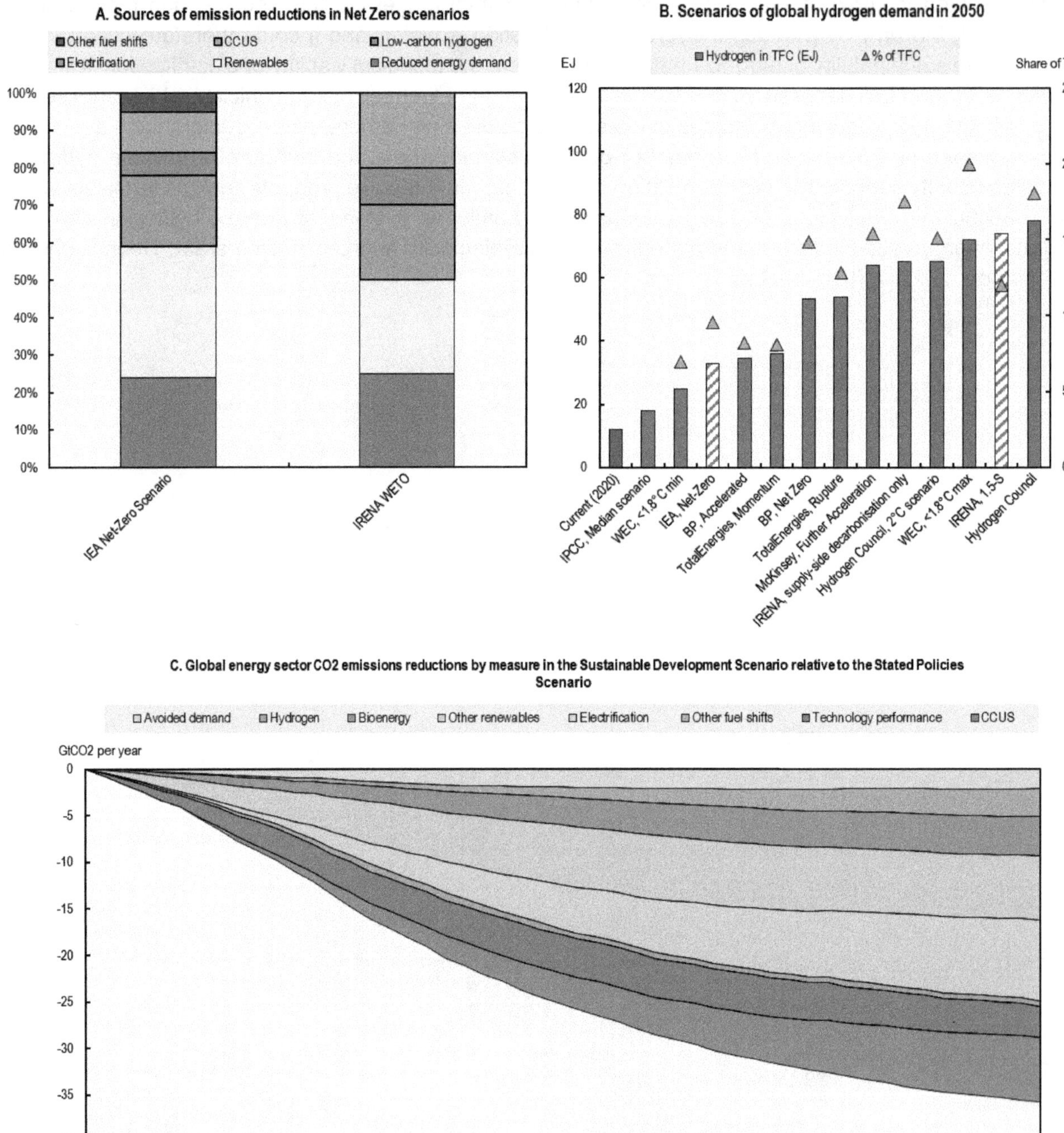

Source: (IEA, 2021[23]), Figure 1.4b OECD analysis based on stated scenarios.

Global production of hydrogen is anticipated to expand significantly in the coming decades as the fuel and its derivatives play an increasing role in achieving Net Zero targets. Projections for the share of renewable hydrogen in global energy demand by 2050 vary between 12 and 22%, equivalent to 340-420 exajoules (EJ) of energy (Figure 1.4 b). Not all of this hydrogen will be 'green', with a notable amount likely to be produced using natural gas and carbon capture and storage (CCUS) technologies. Nevertheless, the increase in hydrogen production required for these scenarios is vast. In 2021, global hydrogen production was equal to 94 megatons (Mt), with over 90% of this being produced using unabated fossil fuels. By 2050, the International Energy Agency (IEA) anticipates that global production will need to reach 500 Mt to meet global demand, with over 90% this coming from low-carbon sources (IEA, 2022c).

Demand for renewable hydrogen will grow in a range of sectors over the coming decades as it is expected to become cost competitive with fossil fuel equivalents due to technology advancements and regulatory push (Figure 1.5). Deep decarbonisation implies that a number of hard-to-abate sectors will need to find novel and innovative ways of integrating low-carbon fuels into their operations. By 2050, industrial demand for hydrogen could reach 62.9Mt, almost double the level of industrial hydrogen consumption in 2020 (IEA, 2023[36]). As the cost of renewable hydrogen comes down, its use in a number of novel applications, such as fuel-cell vehicles, shipping and heavy industry is projected to rise from zero in 2020 to 93.2Mt, 57.3Mt, and 28.3Mt respectively by 2050 (Figure 1.5). The challenge in the shorter-term is that the degree of risk involved in technology switching and the concomitant implication for infrastructure means that potential drivers of demand, such as those in hard-to-abate industries, face high initial costs and limited incentives to adopt low-carbon alternatives.

Figure 1.6. Global hydrogen demand by sector in IEA Net Zero Scenario (2020-2050)

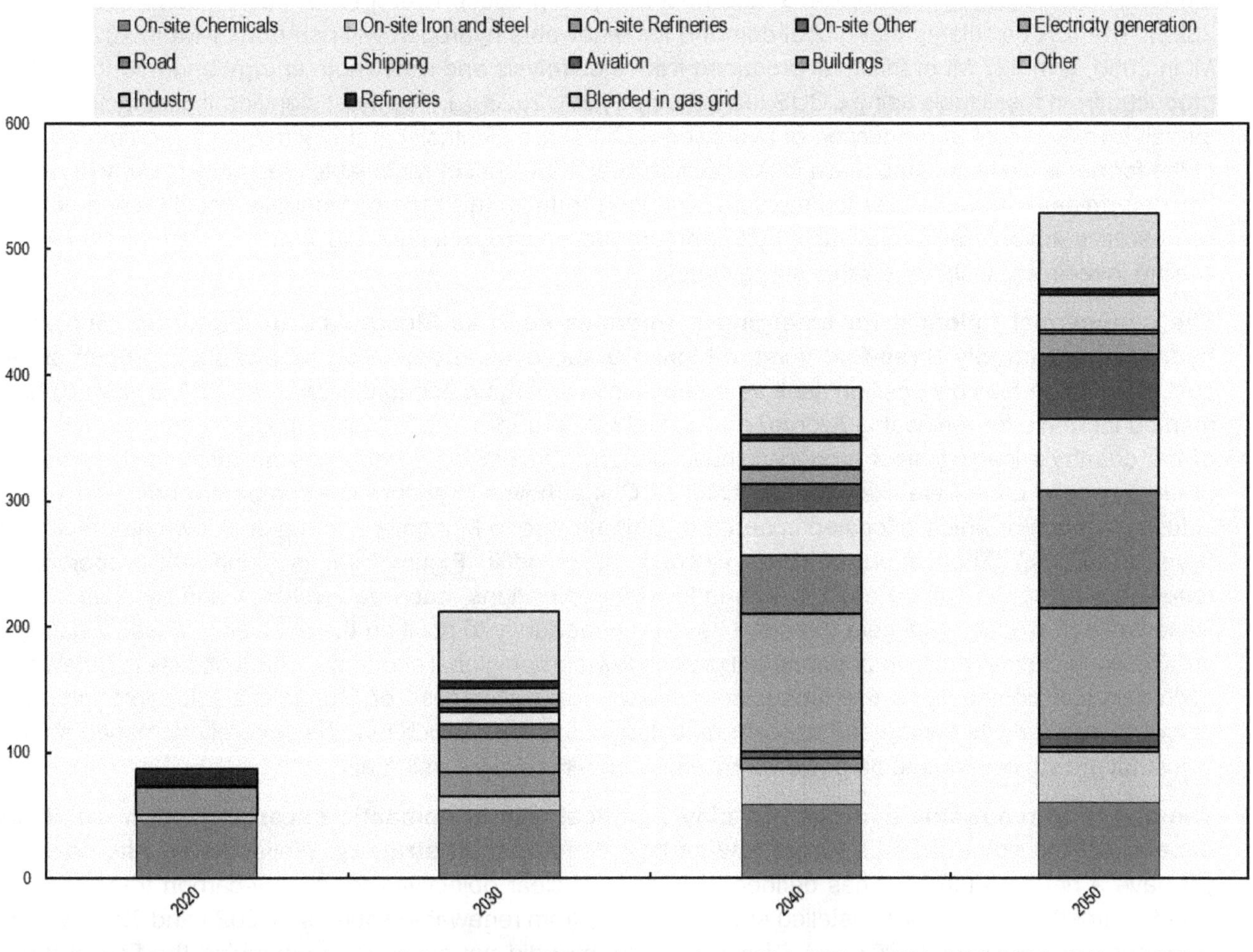

Note: The share of low-carbon hydrogen in total demand is projected to increase from 11% in 2020, to 70% in 2030, 91% in 2040, and 99% in 2050.
Source: (IEA, 2021[23])

Hydrogen is one of the few technologies that can both support the decarbonisation of Mongolia's industrial and energy sectors. Renewable hydrogen has the advantage of being able to support the decarbonisation of Mongolia's emissions-intensive mining sector whilst also contributing to the decarbonisation of the country's fossil fuel intensive power system. In the mining sector, which is likely to serve as the initial source of renewable hydrogen demand within the country, hydrogen can be used in heavy-duty mine vehicles for which electrification is not viable (Nilsson et al., 2021[29]). In the power sector, hydrogen can be an alternative for dispatchable green energy, with hydrogen that is produced from VRE sources offering low-carbon grid balancing solutions and enabling Mongolia in the short- and medium-term to navigate some of the grid inflexibilities that pose technical barriers to wider integration of VREs into the national grid. Other potential applications in the country include the production of synfuels and their use in public transport systems. Nevertheless, it should be noted that the development of a hydrogen industry is not a necessary condition for RES expansion and any new RES capacity for hydrogen production should offer additionality to the grid if it is to be connected.

As the global renewable hydrogen market matures and expands there may be trade opportunities for emerging economies such as Mongolia. While innovation and technology development linked to renewable hydrogen production remains concentrated in a small number of countries (primarily Japan, the United States, Germany and Korea), many emerging economies – including Mongolia – are endowed with the renewable energy resources and land necessary to produce hydrogen at scale (Cammeraat et al., 2022). The IEA calculates that global demand for renewable hydrogen will rise from 1 Mt in 2022 to 420 Mt in 2050, with 327 Mt of this total produced from electrolysis and renewable energy and a further 89 Mt produced from fossil fuels using CCUS technologies (IEA, 2023[36]). The most significant cost factors in the overall levelised cost of production of hydrogen (LCOH) are electricity and electrolysers, and as the cost of the former is likely to come down in the coming years, the cost of renewable electricity is likely to be the key determinant of LCOH over the medium- and long-term. In this regard, Mongolia could have a notable comparative advantage in renewable hydrogen production should it succeed in addressing the barriers to scaling investment in its renewable energy sector.

The commercial rationale for emerging economies such as Mongolia's in exploring renewable hydrogen are largely three-fold. First, if Mongolia succeeds in producing renewable hydrogen at low cost, then it can feasibly position itself as a supplier to emerging economies where there is likely to be a marked increase for renewable hydrogen as a fuel for hard-to-abate industry. In China, for example, half of the country's leading steel producers have begun to invest in hydrogen as an effort to decarbonise production, and others are likely to follow suit as China moves to secure the competitiveness of its steel industry – much of which is located in northern China, close to Mongolia – in a global low-carbon context (Hseuh, 2023[37]) (Zhou et al., 2022[38]). Second, as noted in Figure 1.4a., a significant proportion of renewable hydrogen will be used in non-industrial applications, such as fertilisers and synthetic fuels. Renewable hydrogen producers therefore have an opportunity to position themselves in nascent but high growth value chains whose importance will grow in low-carbon global economy. Third, access to renewable hydrogen can contribute to the future international competitiveness of Mongolia's tradeable industrial output (particularly its mining and nascent metallurgical sector), which may be severely damaged should industrial growth continue to be powered by emissions-intensive fossil-fuels.

The ability of renewable hydrogen to play a critical role in domestic decarbonisation will likely depend on the government's future low-carbon development strategy. Whilst the country does not yet have a net zero target, it has defined a number of clear objectives for its low-carbon transition: the country aimed to have 20% of installed energy capacity from renewable sources by 2023 and 30% by 2030 (Government of Mongolia, 2019[39]). Though the country did not succeed in achieving the first of these milestones, there remains time to meet its 2030 target. The government also set out a 27.2% GHG reduction compared to the business-as-usual (BAU) scenario in its NDC. Nevertheless, the maturity of the country's low-carbon policy framework development remains low, with Mongolia yet to develop strategies for coal phase-out, industrial decarbonisation plans, a mechanism for carbon pricing, nor an overarching strategic roadmap to coordinate the systemic nature of the decarbonisation challenge. The ongoing work of the Ministry of Economy and Development to establish a Long-Term Vision 2050 as a foundation for a future fully-fledged long-term low- emission development strategy (LT-LEDS) can be seen as a step towards the articulation low-carbon technology adoption and development contributions to an economy-wide decarbonisation scenario. This work towards a LT-LEDS is supported by international partners including the 2050 Pathway Platform, the EBRD, the ADB, IDDRI, the OECD and the UNDP.

If developed, Mongolia's vast renewable energy capacities could allow renewable hydrogen to play a role in the country's broader low-carbon development pathway. Mongolia has an estimated 2.6 terawatts (TW) of renewables capacity, with a hypothetical combined electricity output from the country's wind and solar resources potentially reaching 15,000 terawatt-hours (TWh) per year – double the China's total electricity consumption in 2020 (IRENA, 2016[40]) (IEA, 2022[13]). Despite the country's vast renewable energy potential, the actual level of renewable energy supply remains very limited. As of 2020, the share of renewables in total energy supply (TES) was only 2%, though this did represent a 9.6% year-on-year increase and a 40.4% increase over the preceding five years (IRENA, 2022[41]). Shares of electricity generation from renewable sources are concomitantly low, at 10% of total generation. Data from international organisations is generally in line with the government's own statistics, with the Mongolian authorities calculating in 2020 that the country had 649.4GWh of installed renewable electricity capacity (wind 70.4%; solar 16.8%; hydro 12.8%), equal to 9.1% of total installed capacity, with the remaining 90.9% or 4,493.6GWh accounted for by coal-fired CHP plants (Ministry of Energy of Mongolia, 2022[42]).

Mongolia can pursue new, green industrial jobs and innovation if the country succeeds in developing its renewable energy potential. Not only can the development of the country's renewable energy resources help decarbonise the power and industrial sectors, but it can also enable the country to diversify its economy in new green industries where the access to abundant, cheap renewable energy is a key input. Similarly, it can increase the attractiveness of energy-intensive existing industries looking to secure similarly abundant, cheap and sustainable energy. It is in this context that the Government of Mongolia is pursuing low-carbon fuels such as renewable hydrogen, with a number of private sector-led initiatives in Mongolia indicating that investors and entrepreneurs recognise Mongolia as a potential location for a renewable hydrogen industry.

Private investors are already developing projects to produce renewable hydrogen in Mongolia. Reflecting the opportunities for renewable hydrogen production identified in recent reports by organisations such as the New Climate Institute, a number of private investors are already developing renewable hydrogen projects in Mongolia (Nilsson et al., 2021[29]). This report draws on several case studies of these early-stage projects, outlined in Annex A, to present a number of policy-relevant lessons throughout. What emerges in the case studies is that, despite the significant obstacles to getting low-carbon projects off the ground in Mongolia, there is significant interest from both established private sector actors in Mongolia and firms which are new to the country.

Policymakers are also beginning to explore the potential for hydrogen to contribute to a number of policy objectives of the Mongolian government. The establishment of the Mongolian Hydrogen Council in 2022 was the clearest example yet of public-private interest in the development of a renewable hydrogen industry in Mongolia, which is a body that brings together a number of senior public and private sector representatives to discuss opportunities for hydrogen development in Mongolia. Yet, interest also comes from elsewhere. For example, one of the case studies included in this report features a project created under the aegis of the recently established Industrialisation Policy Committee of Mongolia, highlighting the fact that the technology is regarded as an opportunity for new value creation in Mongolia as much as it is for its decarbonisation. Coordinating interested policymakers and line ministries and enabling them to leverage the potential of early-stage, enthusiastic pilot projects will be key for the country to demonstrate the viability of hydrogen as a vector of local decarbonisation and industrial development.

Box 1.2. The different shades of hydrogen

Many international organisations, national administrations and private sector actors use 'colour' classifications to denote the differing production methods for hydrogen. These colours are used to approximate the carbon intensity and sustainability of different approaches to hydrogen production, and an overview of these 'classifications' is as follows:

- **Renewable hydrogen** is hydrogen produced using electrolysers powered by renewable energy sources. As with all forms of hydrogen it can be burned as a fuel and used as a feedstock in other products; in the case of the former it is carbon neutral as the only by product is water, in the case of the latter it can contribute to low-carbon or zero carbon products and hydrogen derivatives.
- **Blue hydrogen** is renewable hydrogen produced from natural gas but using carbon capture to limit emissions. Critics of blue hydrogen argue that investments in the technology may slow the transition to carbon neutrality as it could compete with the commercialisation of renewable hydrogen. Nevertheless, it has been proposed as a short-term option to support the decarbonisation of existing grey hydrogen production and facilitating the emergence of a more mature hydrogen market.
- **Grey hydrogen** is hydrogen produced using energy from natural gas using steam reforming, without mitigation technologies.
- **Purple hydrogen** is hydrogen produced using nuclear energy.
- **Brown or black hydrogen** is hydrogen produced from coal.

At present, over 90% of global hydrogen production is powered by fossil fuels. This is in part a reflection of the **maturity of the technology used in fossil fuel-based production**, a corollary of which being the **low level of technological maturity and concomitantly high cost of 'green' hydrogen production**. Most national strategies for the development of renewable hydrogen aim to lower the cost of production and consumption of renewable hydrogen versus its fossil fuel-produced equivalents, in doing so **facilitating the emergence of renewable hydrogen markets** and improving its commercial viability as a low-carbon fuel and feedstock.

There has been **criticism on the use of 'colours' to denote hydrogen production** methods. It has been argued that the use of colours to refer to different production routes, as well as unqualified terms such as 'sustainable', 'low-carbon' or 'clean' can obscure different levels of associated emissions. In addition, a **lack of standardisation across the terminology risks creating impracticalities for contracting and investment decisions**. Nevertheless, while recognising that there presently lacks an international agreement on the use of differing terminologies for hydrogen production methods, this report uses 'renewable hydrogen' to denote hydrogen produced using electrolysis from renewable energy sources and 'low-carbon' to denote hydrogen produced using different combinations of renewable technologies and carbon mitigation technologies.

Source: Adapted from (Cammeraat et al., 2022; International Energy Agency, 2023a, 2023b, 2021; OECD, 2022)

A renewable hydrogen development strategy could support Mongolia's clean energy transition as well as domestic industrial decarbonisation

The literature on policies to accelerate investment in clean hydrogen (and, in particular, private investment) highlights the high level of risk of the sector. On the one hand, clean hydrogen projects require substantial capital investment for the production (electrolysers and, in most projects, renewable energy production equipment), and for the transportation and storage infrastructure. On the other hand, the sector is nascent and facing high uncertainties in a great number of areas that interact with one another (systemic risk): location and availability of supply and demand, availability of infrastructure, selling prices, regulations, etc. This is why, in countries with strong interest for renewable hydrogen consumption or production, policymakers are looking at all available options for providing some level of certainty and decreasing the perception of risk. Over the past few years, many governments have been issuing national renewable hydrogen strategies with this intent. For example, the European Union, Japan and the Republic of Korea have been setting import targets, which combined reach nearly 16 Mt by 2030 (OECD and the World Bank, Forthcoming[43]), thus providing a visibility on the future clean hydrogen market for exporters.

Even though all governments face similar challenges with regards to the level of risk, emerging economies encounter a set of additional, specific barriers to the development of renewable hydrogen and low-carbon technologies in general. The challenges of how to achieve the type of systemic transformation necessary for achieving net zero are shared by advanced and emerging economies alike: policymakers across the OECD, just as in Mongolia, grapple with how to overcome the market failures that prevent investment and innovation being directed to areas that will lower global emissions and create new, more sustainable economic growth. Nevertheless, the starting point is not necessarily the same, and the hurdles in achieving low-carbon growth for emerging and developing economies can be higher than in more advanced equivalents.

- **Public finance**: The governments of many emerging and developing economies often face considerably more challenging public finance constraints than those in the OECD. The cost of government borrowing can be higher, making it more onerous to finance the type of large-scale public-backed infrastructure and technology programmes that characterise the approach of many more developed economies to the low-carbon transition. Recent work by the IEA suggests, in the current high interest rate context, that for project developers in an emerging markets an increase in the weighted average cost of capital (WACC) from 6% to 15% would result in a 70% increase in the LCOH of hydrogen produced from solar (IEA, 2023[44]). This means the governments of emerging economies are often required to do more with less, or to turn to international financial institutions such as the IMF or World Bank to bridge the gap and de-risk necessary infrastructure investments. In a context of constrained public finance, success with technology and infrastructure choices can matter significantly more than in a developed economy. Unlike developed economies, where both the public and private sector may be more capable of pursuing multiple strategies and avenues then choosing the most promising option at a later point, emerging economies may not enjoy this luxury. Therefore, decisions to deviate from a technologically open approach to low-carbon development to one that is more directional must be carefully weighted with the potential risks.
- **Political economy of extractive industries in development contexts**: In many extractive-based emerging economies, direct or indirect rents from the very extractive sectors that are liable for significant disruption in the context of the low-carbon transition are essential for government revenues. How the revenues from extractive industries are used whilst those industries are still internationally competitive can have a significant impact on the inclusivity of the low-carbon transition. In some countries, such as Norway and other oil-rich economies in the MENA region, governments are actively using revenues from the oil and gas sector to support low-carbon industrial development and innovation, paving the way for diversified and sustainable growth in a

post-fossil fuel context. The challenge of redirecting extractive sector revenues and policy support towards low-carbon development is as much one of finance as it is of political economy.

- **Investment mobilisation**: In the absence of sufficiently developed local capital markets, emerging and developing economies often place a great deal of importance on attracting foreign direct investment (FDI), which creates opportunities to spur industrialisation, induce technology transfer, and create high-quality and productive jobs. Yet, emerging and developing economies already face challenges in attracting high quality investment. Issues in the broader business climate, challenges around regulatory implementation, governance and transparency issues are among the many factors that raise the risk associated with many emerging economies over more advanced equivalents. In the period of low interest rates that characterised the years following the global financial crisis, the higher risk associated with emerging markets made them more attractive options for some investors looking for a higher rate of return. Now, in a global financial context that is much changed and where interest rates on even high-grade government debt produce significant yields, there is a risk that the difficulty faced by emerging economies in attracting foreign direct investment will be redoubled. In the context of financing infrastructure, industry, and technology for the low-carbon transition, where the investment needs across many sectors of the economy are substantial, there is an implicitly higher level of risk and uncertainty from the outset, further compounding the difficulties facing emerging and developing economies.
- **Innovation and technology capacities**: Firms in emerging and developing economies may face issues of technological and innovation agency in the low-carbon transition. In the World Intellectual Property Organisation's ranking of the most innovative economies according to the Global Innovation Index, the 35 most innovative countries in the world are all high-income, with the notable exception of China (World Intellectual Property Organization, 2023[45]). Green innovation follows a similar pattern: high-income countries have better innovation environments for a low-carbon future than their low-income peers, with wealth a significant determinant of the ability of a country to define its own low-carbon future (MIT Technology Review, 2023[46]). This means that where firms may be required or incentivised to adopt new low-carbon technologies, fuels, and processes, in many cases the innovation that drives these changes will originate elsewhere. Governments of emerging and developing economies therefore face a challenge in ensuring that their firms and workers are not just shaped by external developments but can have agency in the innovation and direction of such technologies and in so doing increase their opportunities for benefiting from the global low-carbon transition.
- **Competitiveness**: In the short- to medium-term, the low-carbon transition may negatively impact the competitiveness of emerging and developing economies to an extent that is more severe than in more developed countries. Firms in emerging economies already face significant competitiveness challenges: limited investment capacities at the firm-level can make it hard to be innovative, the high cost or limited access to finance can create onerous burdens on firms, and, particularly in the context of Eurasia, competition issues around state ownership and the governance of state-owned enterprises can complicate firm activities and limit their opportunities. All these factors risk being aggravated in a context where firms are asked to switch to fuels and technologies that in the short-term are more costly, or where firms have only recently invested in capital intensive assets with a long lifespan that risk being stranded.

In Mongolia, the economic challenges faced by other emerging economies are exacerbated by geography and market structure. Mongolia is a vast, landlocked country. As such, its economy already faces a severe internal and external connectivity penalty: it is harder, longer, and more expensive for a Mongolian firm to deliver its products to the market. Any policy that risks a further detrimental impact on the competitiveness of Mongolia's private sector in the short-term, even if in the long-term it may be essential to their competitiveness, will have to be carefully designed to minimise its disruptiveness. At the same time, the predominance of the mining industry in Mongolia's economy may also indirectly curtail the competitiveness of the country's private sector. For example, through the so-called 'Dutch Disease', whereby the competitiveness of tradeable non-extractive sectors in extractive-based economies are negatively affected through exchange rate appreciation, structural barriers to diversification are created.

A national strategy can help address the challenges and make the most of new opportunities in developing a renewable hydrogen industry. National renewable hydrogen strategies can help accelerate the development of a renewable hydrogen industry as well as ensuring that it plays the desired role in achieving other government policy objectives, such as in industry development and decarbonisation. Strategies can help send signals to investors that the government is willing to support in establishing a framework that enables hydrogen projects to succeed and can align the necessary government and private sector stakeholders necessary to overcome policy, regulatory, technological, and social barriers. This may be particularly true when such technology-specific strategies are formulated in parallel with strategies for the broader, systemic transition necessary for low-carbon and sustainable economic growth.

Mongolia's rationale for supporting the domestic production of renewable hydrogen captures the dual sustainability and industrial rationale of many governments. In its 'New Recovery Policy', a ten-year economic development plan issued by the Government of Mongolia in late 2019, the government highlighted the development of renewable hydrogen production in Mongolia as a strategic objective, notably in the South Gobi region. This marked the first time the government had identified renewable hydrogen production as an area of national interest, building on broader commitments to the development of renewable energy production and environmental sustainability set out in its Vision 2050 and NDC. As mentioned previously, the vast renewables potential of Mongolia has also contributed to attracting a number of pilot renewable hydrogen projects to Mongolia.

The development of a renewable hydrogen strategy can help envision a way for Mongolia to take advantage of current opportunities for moving forward the national agenda towards a lower-carbon and sustainable economic growth. National stakeholders find that Mongolia has a number of assets which could make it a suitable candidate for renewable hydrogen production, and possibly exports. Firstly, the country is well endowed with at least one of the major inputs for the production of renewable hydrogen – renewable energy generation potential– and faces relatively few competing land use claims. Secondly, while the Mongolian market for renewable hydrogen may be relatively small, the country is proximate to a number of markets where demand for renewable hydrogen is likely to grow significantly in the coming years, notably China but also Japan and South Korea. Harnessing the potential for hydrogen production at competitive cost thus appears technically feasible, but this will require a significant degree of policy intervention and coordination to overcome challenges, as explored in this report. Building on a public-private dialogue with investors pursuing low-carbon innovation in Mongolia, developing a national renewable hydrogen strategy is a good opportunity to define the potential of the country, identify avenues for the implementation of a sound and realistic plan for realising this potential, and defining the policies that can underpin implementation.

References

ADB (2024), *Energy Policy*, ADB, https://www.adb.org/what-we-do/energy-policy. [32]

Amoroso, S. et al. (2021), *World Corporate Top R&D Investors: Paving the way for climate neutrality, A joint JRC and OECD report*, Publications Office of the European Union, Luxembourg, https://web-archive.oecd.org/2021-11-22/617125-world-corporate-top-rd-investors-paving-the-way-for-climate-neutrality.pdf. [58]

BP (2022), *BP Statistical Review of World Energy*, BP, https://www.bp.com/content/dam/bp/business-sites/en/global/corporate/pdfs/energy-economics/statistical-review/bp-stats-review-2022-full-report.pdf. [5]

Brian S. Fisher, T. (2011), *The development of the Oyu Tolgoi copper mine: an assessment of the macroeconomic consequences for Mongolia*, BAEconomics Pty Ltd. [11]

Cammeraat, E., A. Dechezlepretre and G. Lalanne (2022), *Innovation and Industrial Policies for Green Hydrogen*, OECD, Paris, https://www.oecd-ilibrary.org/docserver/f0bb5d8c-en.pdf?expires=1692286715&id=id&accname=guest&checksum=767FA1D7985F43131BCEA25FFAC2755B. [35]

Criscuolo, C. et al. (2022), *An industrial policy framework for OECD countries: old debates, new perspectives*, OECD Science, Technology and Industry Policy Papers, No. No. 127, OECD, Paris, https://www.oecd-ilibrary.org/docserver/0002217c-en.pdf?expires=1692268763&id=id&accname=ocid84004878&checksum=1E15ADDA4BEDB5A445495656933FF0E9. [34]

Dostal, J. and O. Gerel (2023), *Rare Earth Element Deposits in Mongolia*, Minerals, No. 13(1), https://doi.org/10.3390/min13010129. [59]

European Commission (2024), *Carbon Border Adjustment Mechanism*, European Commission, https://taxation-customs.ec.europa.eu/carbon-border-adjustment-mechanism_en. [56]

Government of Mongolia (2023), *New Recovery Policy*, Government of Mongolia, https://www.nrpa.gov.mn/en/new-recovery-policy. [17]

Government of Mongolia (2020), *"Vision 2050" Long-Term Development Policy of Mongolia*, Government of Mongolia, https://cabinet.gov.mn/wp-content/uploads/2050_VISION_LONG-TERM-DEVELOPMENT-POLICY.pdf. [16]

Government of Mongolia (2019), *Mongolia's Nationally Determiend Contribution to the United Nations Framework Convention onf Climate Change*, Government of Mongolia, http://www.en.ccrcc.mn/stat/file/u/NDC_1.pdf. [26]

Government of Mongolia (2019), *Mongolia's Nationally Determined Contribution to the United Nations Framework Convention on Climate Change*, UNFCCC, New York, https://unfccc.int/sites/default/files/NDC/2022-06/First%20Submission%20of%20Mongolia%27s%20NDC.pdf. [39]

Government of Mongolia (2015), *Intended Nationally Determined Contribution of Mongolia*, Government of Mongolia, http://www.ccrcc.mn/stat/file/u/INDC_Full-Document_ENG_2016.pdf. [27]

Greenhouse Gas Protocol/World Resources Institute and World Business Council for Sustainable Development (2004), *The Greenhouse Gas Protocol: A Corporate Accounting and Reporting Standard*, Greenhouse Gas Protocol/World Resources Institute and World Business Council for Sustainable Development, https://ghgprotocol.org/sites/default/files/standards/ghg-protocol-revised.pdf. [55]

Gutschow, J. et al. (2022), *The PRIMAP-hist national historical emissions time series*, Earth Syst. Sci., No. Data, 8, pp. 571-603, https://doi.org/10.5194/essd-8-571-2016. [14]

Hseuh, C. (2023), *The Chinese Steel Industry: Three Pathways to Green Steel*, China Brief, No. Vol. 23 Issue 5, Jamestown Foundation. [37]

IEA (2023), *Coal in Net Zero Transitions: Strategies for rapid, secure and people-centred change*, IEA, Paris, https://iea.blob.core.windows.net/assets/4192696b-6518-4cfc-bb34-acc9312bf4b2/CoalinNetZeroTransitions.pdf. [25]

IEA (2023), *Global Hydrogen Review*, IEA, Paris, https://iea.blob.core.windows.net/assets/8d434960-a85c-4c02-ad96-77794aaa175d/GlobalHydrogenReview2023.pdf. [44]

IEA (2023), *Greenhouse Gas Emissions from Energy (database)*, https://www.iea.org/data-and-statistics/data-product/greenhouse-gas-emissions-from-energy. [52]

IEA (2023), *Mongolia*, Greenhouse Gas Emissions from Energy, IEA, Paris, https://www.iea.org/countries/mongolia. [21]

IEA (2023), *Mongolia*, Energy and Emissions per Value Added Database, IEA, Paris, https://www.iea.org/data-and-statistics/data-product/energy-and-emissions-per-value-added-database. [15]

IEA (2023), *Net Zero Roadmap: A Global Pathway to Keep the 1.5C Goal in Reach - 2023*, IEA, Paris, https://iea.blob.core.windows.net/assets/7c02e774-9d1b-4398-9313-840913e1b4e6/NetZeroRoadmap_AGlobalPathwaytoKeepthe1.5CGoalinReach-2023Update.pdf. [36]

IEA (2022), *World Energy Balances*, https://www.iea.org/countries/mongolia. [13]

IEA (2021), *Net Zero by 2050: A Roadmap for the Global Energy Sector*, IEA, Paris, https://iea.blob.core.windows.net/assets/deebef5d-0c34-4539-9d0c-10b13d840027/NetZeroby2050-ARoadmapfortheGlobalEnergySector_CORR.pdf. [23]

IEA (2021), *Phasing Out Unabated Coal: Current Status and Three Case Studies*, IEA, Paris, https://iea.blob.core.windows.net/assets/861dc94d-a684-4875-80fb-a1faaf914125/PhasingOutUnabatedCoal-CurrentStatusandThreeCaseStudies.pdf. [53]

IEA (2018), *Mongolia Renewable Energy Feed-in Tariff*, IEA, Paris, https://www.iea.org/policies/6469-mongolia-renewable-energy-feed-in-tariff. [57]

IMF (2023), *World Economic Outlook Database, April*, https://www.imf.org/en/Publications/WEO/weo-database/2023/April. [1]

IMF (2022), *Annex 1.SF.1. Coal Dependence: Econometric Analysis and Energy Transition Risks for Coal Producers*, IMF World Economic Outlook, IMF. [6]

IMF (2021), *Energy Subsidy Template*, IMF, DC, https://www.imf.org/-/media/Files/Topics/Environment/energy-subsidies/fuel-subsidies-template-2022.ashx. [47]

IRENA (2022), *Energy Profile: Mongolia*, IRENA, https://www.irena.org/-/media/Files/IRENA/Agency/Statistics/Statistical_Profiles/Asia/Mongolia_Asia_RE_SP.pdf?rev=043f5c4c85054e179be1ae1b397e5972. [41]

IRENA (2020), *Green hydrogen cost reduction*, https://www.irena.org/publications/2020/Dec/Green-hydrogen-cost-reduction. [51]

IRENA (2016), *Renewable Readiness Assessment: Mongolia*, IRENA, https://www.irena.org/-/media/Files/IRENA/Agency/Publication/2016/IRENA_RRA_Mongolia_2016.pdf?rev=b7b8dc32d10745cc9b70b55d31569366. [40]

Ministry of Energy of Mongolia (2022), *Head Supply and Renewable Energy Policy Planning*, Ministry of Energy of Mongolia, Ulaanbaatar, https://www.irena.org/-/media/Files/IRENA/Agency/Events/2022/May/Heat-supply-and-renewable-energy-policy-planning.pdf?la=en&hash=FC805CF764BBD7F6A001D27B9FDE9FF4170A0B48. [42]

Ministry of Energy of Mongolia (2022), *Heat Supply and Renewable Energy Policy Planning*, Ministry of Energy of Mongolia, Ulaanbaatar, https://www.irena.org/-/media/Files/IRENA/Agency/Events/2022/May/Heat-supply-and-renewable-energy-policy-planning.pdf?la=en&hash=FC805CF764BBD7F6A001D27B9FDE9FF4170A0B48. [28]

MIT Technology Review (2023), *The Green Future Index*. [46]

National Statistical Committee of Mongolia (2022), *Contribution of sectors to changes in GDP*, National Statistical Committee of Mongolia. [8]

National Statistics Office of Mongolia (2023), *Foreign Trade*, https://www.1212.mn/en/statistic/statcate/573062/table/573062. [49]

National Statistics Office of Mongolia (2023), *Investment*, https://www.1212.mn/en/statistic/statcate/573075/table/573075. [48]

National Statistics Office of Mongolia (2023), *National Accounts*, https://www.1212.mn/en/statistic/statcate/573052/table-view/DT_NSO_0500_001V1. [9]

Nilsson, A. et al. (2021), *Green Hydrogen Applications in Mongolia: Technology potential and policy options*, New Climate Institute, https://newclimate.org/sites/default/files/2021/10/NewClimate_Green_Hydrogen_Applications_in_Mongolia.pdf. [29]

OEC (2023), *Mongolia*, OEC, https://oec.world/en/profile/country/mng?depthSelector1=HS6Depth&tradeScaleSelector1=tradeScale2. [4]

OECD (2023), *Air pollution exposure*, OECD, Paris, https://www.oecd-ilibrary.org/environment/air-pollution-exposure/indicator/english_8d9dcc33-en. [22]

OECD (2023), *Green Growth Indicators*, OECD Publishing, Paris, https://stats.oecd.org/Index.aspx?DataSetCode=GREEN_GROWTH. [20]

OECD (2023), *Mongolia*, OECD Green Growth Indicators, OECD, Paris, https://stats.oecd.org/Index.aspx?DataSetCode=GREEN_GROWTH. [19]

OECD (2022), *Equitable Framework and Finance for Extractive-based Countries in Transition (EFFECT)*, OECD, Paris, https://www.oecd-ilibrary.org/docserver/7871c0ad-en.pdf?expires=1694092451&id=id&accname=ocid84004878&checksum=CDADE16BB122E467FE2BD3B6036DDB69. [24]

OECD (2022), *Green Hydrogen Opportunities for Emerging and Developing Economies*, OECD, Paris, https://www.oecd-ilibrary.org/docserver/53ad9f22-en.pdf?expires=1696414509&id=id&accname=ocid84004878&checksum=01FB008F220DF39E4DDDC822F7FFEB07. [54]

OECD (2021), *Agreement reached at OECD to end export credit support for unabated coal-fired power plants*, OECD, Paris, https://www.oecd.org/newsroom/agreement-reached-at-oecd-to-end-export-credit-support-for-unabated-coal-fired-power-plants.htm. [31]

OECD (2021), *Trade in Value Added (TiVA) 2021 ed: Principal Indicators*, OECD Publishing, Paris, https://stats.oecd.org/Index.aspx?DataSetCode=TIVA_2021_C1. [12]

OECD and the World Bank (Forthcoming), *Global Cloean Hydrogen Financing Publication*. [43]

U.S. Department of State (2023), *The United States and Mongolia Sign MOU to Collaborate on Critical Minerals*, U.S. Department of State, https://www.state.gov/the-united-states-and-mongolia-sign-mou-to-collaborate-on-critical-minerals/. [50]

UNFCCC (2023), *COP28 Agreement Signals "Beginning of the End" of the Fossil Fuel Era*, UNFCCC, https://unfccc.int/news/cop28-agreement-signals-beginning-of-the-end-of-the-fossil-fuel-era. [30]

UNFCCC (2023), *Mongolia's National Inventory Report 2023*, UNFCCC, https://unfccc.int/sites/default/files/resource/20231112_NIR_MGL.pdf. [18]

World Bank (2023), *Mongolia Economic Update: Partial Recovery and Lingering Risk*, World Bank, DC, https://thedocs.worldbank.org/en/doc/97b959e7c02bfd04f4ab3c1bdd5b9e20-0070012023/original/Mongolia-Economic-Update-April-2023.pdf?fbclid=IwAR1vdhhH9PCW7UbprWMrUnJNg6osVGY5JpU38b61cbAlVwnmXnpYKeRUIQw. [7]

World Bank (2023), *Mongolia Economic Updated April 2023*, World Bank, DC, https://documents1.worldbank.org/curated/en/099413204212338504/pdf/IDU0c9bb5b690643104ccf091520b7f0565236e4.pdf. [10]

World Bank (2023), *Poverty and Inequality Platform*, World Bank, DC, https://pip.worldbank.org/poverty-calculator?src=MNG. [2]

World Bank (2023), *Scaling Up to Phase Down: Financing Energy Transition in Developing Countries*, World Bank, https://www.worldbank.org/en/news/press-release/2023/04/20/scaling-up-to-phase-down-financing-energy-transition-in-developing-countries. [33]

World Bank (2023), *World Development Indicators*, World Bank, DC, https://databank.worldbank.org/source/world-development-indicators. [3]

World Intellectual Property Organization (2023), *Global Innovation Index 2023: Innovation in the face of uncertainty*, https://doi.org/10.34667/tind.48220. [45]

Zhou, T. et al. (2022), *China's green steel plans: Near-term policy challenges and Australia-China links in decarbonisation*, Australian National University, https://iceds.anu.edu.au/files/ANU%20policy%20brief%20-%20China's%20green%20steel%20plans_1.pdf. [38]

2 Mobilising the Mongolian science, technology and innovation system for renewable hydrogen development

This chapter examines Mongolia's science, technology, and innovation (STI) framework and identifies ways to leverage it in support of renewable hydrogen development. There is much discussion around the role of green industrial policy in driving down the costs of technologies and rapidly scaling-up the production and use of renewable hydrogen at the global scale. At the same time, hydrogen innovation is primarily concentrated in a few high-income economies, with China being the notable exception. This means that establishing the conditions for effective technology transfer through international mechanisms will play a key role in an emerging economy such as Mongolia. Fostering demand for low-carbon technologies, for example through green public procurement, should also be a priority. Finally, Mongolia should consider how to mobilise existing public resources for innovation to support low-carbon technology and renewable hydrogen development, and address the misalignment of its STI framework with its climate objectives.

The role of innovation policies in national strategies for renewable hydrogen development

Innovation in hydrogen production technologies is needed to lower the costs of producing renewable hydrogen and enable the technology to support net-zero objectives by mid-century. Recent OECD analysis focusing on OECD hydrogen-producing countries identifies five innovation and industrial policy priorities to support the development of renewable hydrogen and accelerate its deployment (Cammeraat, Dechezlepretre and Lalanne, 2022[1]). First, given the low level of technological maturity in renewable hydrogen production and applications, governments can provide more directed research and development (R&D) support, rather than relying on horizontal, technology-neutral instruments such as R&D tax credits. Second, in countries where renewable hydrogen is to be powered by renewable energy from the national grid, it is imperative that there is an efficient supply of renewable energy, and that policy supports continued cost reductions in renewable energy production. Third, as with other low-carbon solutions, the competitiveness of green hydrogen will depend in large part on establishing clear carbon-price trajectories, unwinding fossil fuel subsidies, and providing subsidies for deployment. Fourth, as with many frontier technologies, investors will seek regulatory clarity around green hydrogen, with the role of internationally agreed standards and certification of particular importance as international hydrogen markets begin to mature. Fifth, governments may consider supporting blue hydrogen as an interim solution to facilitate a transition to green hydrogen, provided this is achieved through clearly time-bound short-term mechanisms.

While the lessons from this work have relevance to Mongolia's own renewable hydrogen ambitions, the country faces a number of additional innovation and industrial policy challenges that are specific to an emerging economy context. These challenges include the more limited capacities – technological and financial – of emerging markets to be first movers in hydrogen innovation and deployment, even if they are endowed with promising natural resources (OECD, 2022[2]). Mongolia, furthermore, is not a (grey) hydrogen-producing country and has a small industrial sector which limits the current options for hydrogen applications in a domestic context. As discussed in complementary OECD work, one possible approach for emerging economies to overcome capacity constraints is to take a more targeted approach to renewable hydrogen development wherein emerging economies leverage particular stages of value chains where their national characteristics might create a comparative advantage in the production of renewable hydrogen and its use in industrial applications (Cordonnier and Saygin, 2023[3]). Technology transfer and policies stimulating domestic demand, in this respect, play a key role, and are prominent in national hydrogen strategies of countries seeking to establish themselves as renewable hydrogen producers.

This chapter focusses on innovation and industrial policies that Mongolia can consider in the framework of developing a national renewable hydrogen strategy. The chapter will also explore available solutions to Mongolia for promoting low-carbon technologies at large, including – but not limited to – renewable hydrogen. Section 2.2 begins with an overview of technology transfer in the context of low-carbon industrial development, wherein many of the observations are as true for hydrogen as they are for other low-carbon technologies. Section 2.3 discusses the role of demand-side policies for stimulating and guiding innovation capacities in Mongolia. These policies aim to help the government's renewable hydrogen objectives, while also ensuring the development of capacities for innovation and adoption of other low-carbon technologies. Section 2.4 concludes with an assessment of the current science, technology, and innovation (STI) policy and regulatory framework in Mongolia, its conduciveness to renewable hydrogen development, and some observations on recent hydrogen-related innovation activities in the country.

Channelling renewable hydrogen innovation to Mongolia: the importance of technology transfer

The transition to a low-carbon economy requires systemic change, involving a range of different policy areas – from energy and industry to transport and finance – which taken together set the framework and incentives for the private sector to adopt less emission-intensive modes of production and organisation. Renewable hydrogen is likely to be one solution among many used by countries around the world as they decarbonise different areas of their economies, and there remains significant scope for emissions reductions through interventions at the lower end of the marginal abatement cost curve. Like other low-carbon technologies, the development and diffusion of renewable hydrogen require significant policy coordination and a strategic vision of the role it should play in a given economy. Notably, this includes focussing on green domestic innovation, and effectively coordinating both domestic and international innovation resources in the pursuit of hydrogen development. Coordinating these resources has emerged as a major component of national hydrogen strategies (Box 2.1). It is important that policymakers have the capacities and knowledge necessary to take a systemic view of the whole value chain now and how it might develop in the future. For example, the potential role of renewable hydrogen to substitute grey hydrogen as a feedstock for fertilisers, or completely novel applications such as the integration of renewable hydrogen into mining and mineral value chains to create low-carbon products – in a way that individual private actors might be unable to do. This is typically the vision that countries develop in their national sectoral strategies, including for renewable hydrogen development.

Policies that give direction to investment and innovation are likely to be important for the development of technological competencies in emerging hydrogen value chains. Such policies come from within STI policymaking as well as from other policy domains (trade, energy, investment, taxation, etc.). From an innovation policy perspective, there are a range of levers that policymakers can use to support priority technologies or research areas, particularly when these remain at an early and uncompetitive stage. Directed public R&D is one important lever, but so too are conditional R&D tax credits and other incentives that support market-led private innovation. Importantly, such interventions can be focussed either on a specific technology (hydrogen) or a purpose or goal (such as lowering emissions from a specific sector). In order to accelerate the deployment of low-carbon technologies such as renewable hydrogen, governments are also returning to more traditional industrial policy levers – itself an indication of the need for coordination between different policy domains in advancing a low-carbon industrial strategy. Such policies may be more focussed on unlocking the finance and investment necessary to scale up promising innovations; building and de-risking the enabling infrastructure; and stimulating demand through targeted demand-side and regulatory policy interventions.

Box 2.1. R&D and innovation policies in national renewable hydrogen strategies

1. Most national renewable hydrogen strategies include important innovation and R&D policies and initiatives, though approaches can vary significantly. R&D is a key component in the deployment of the low-carbon hydrogen technology. Specific, targeted R&D tools are needed in countries positioning themselves as major future renewable hydrogen exporters, because horizontal R&D policies will tend to benefit more mature technologies and thus are not the most effective for nascent, far from market technologies (Cammeraat, Dechezleprêtre and Lalanne, 2022[4]).

- **The Moroccan GreenH2A research platform aims to be a national and regional reference for R&D and innovation**. The research mainly focuses on renewable hydrogen, green ammonia and synthetic fuels, within the broader scope of "Power to X" Initiative. According to the strategy, the main objective of the centre is to promote the use of low-carbon molecules in the industrial sector as well as the production of fertilizers (Government of Morocco, 2021[5]).
- **The strategy of Namibia envisions the set-up of a research centre, the Namibia Green Hydrogen Research Institute (NGHRI).** In addition to its focus on R&D, the centre also aims to be a meeting point for university and industry as well as for government consortia to conduct research for renewable hydrogen development. Lastly, the NGHRI has a capacity building mission and plans to offer formal degree programmes such as Masters and PhDs as well as re-skilling and re-training programmes (Government of Namibia, 2022[6]).
- **The Chilean government created a public fund up to USD 260 million for the setup of the Clean Technologies Institute.** The "*Istituto de Tecnologías Limpias*" (ITL) is an innovation platform for clean energy and mining technologies (Government of Chile, 2021[7]). The funding was awarded to a consortium bringing together several Chilean Universities as well as local public companies such as Industrial Association of Antofagasta. The Institute will focus on sustainable mining issues and building a renewable hydrogen value chain (ASDIT, 2023[8]).
- **The Moroccan Cluster is the first renewable hydrogen cluster in Africa**. The Cluster's main goal is to organize the green hydrogen value chain at national level. Hence, its scope is not limited to advance on R&D but also to support public policy drafting in the realm of green derivatives, to contribute to the development of an industrial value chain and to foster job creation through training and re-skilling courses (Government of Morocco, 2021[5]).

In the OECD, a major focus of renewable hydrogen technology policy support aims to accelerate its commercialisation by reducing production costs. The goal is to make renewable hydrogen and its derivatives competitive with 'grey' hydrogen products. The key determinant of low hydrogen production scaling to levels envisaged in a number of Net Zero scenarios will be lowering the costs involved with its production and adoption. In part this is relative: the competitiveness of hydrogen use will depend in large part on the cost of alternative fuels, including those that are fossil fuel based. At present in Mongolia there has been only limited attention paid by the government to how to phase-out fossil fuels, and a more serious engagement with this issue is likely to be necessary as the government advances its renewable hydrogen ambitions. The reduction of more emissions-intensive alternatives will be achieved through regulation and a more systemic use of market-based instruments such as carbon pricing, as discussed below. Part of the solution involves making significant advances to lower the costs of electrolysers and the necessary feedstock of renewable energy (Figure 2.1), which are the two largest factors in reducing renewable hydrogen production costs. As discussed in Chapter 3, addressing the investment barriers to developing Mongolia's vast renewable energy resources will be a key consideration for any future national hydrogen strategy. Much can already be done through the trade and fiscal policy frameworks to de-risk and lower the associated capital costs.

Figure 2.1. Key drivers for lowering green hydrogen production costs

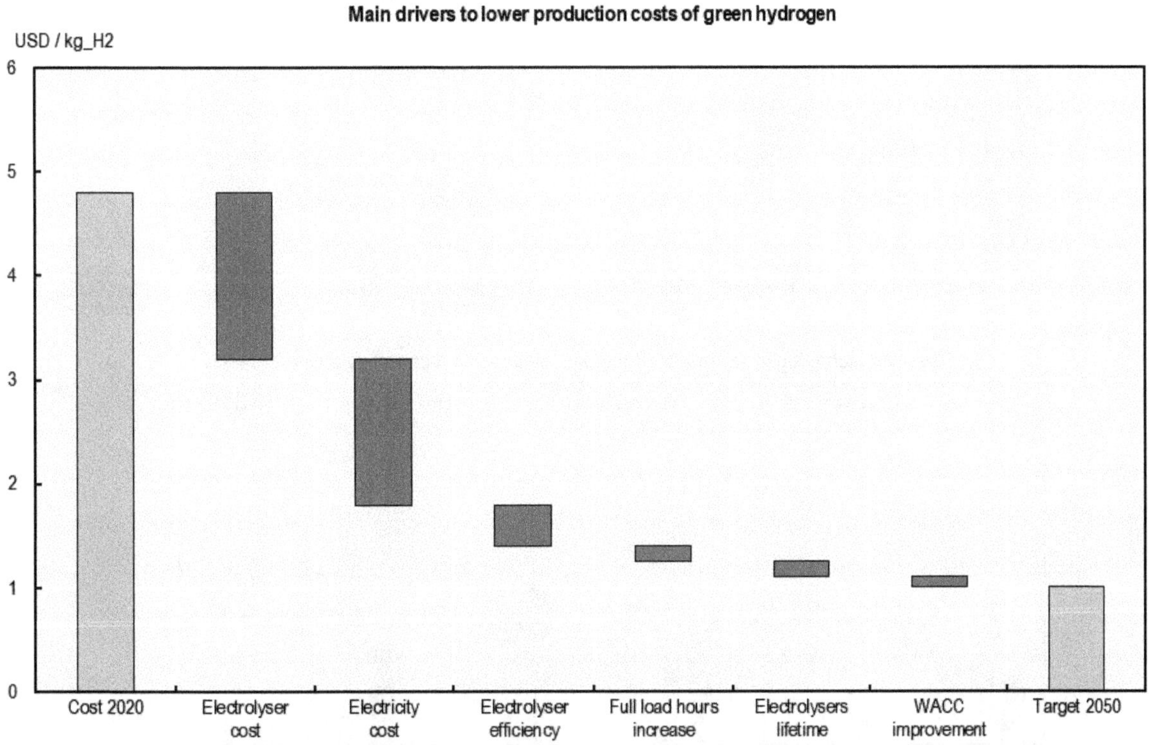

Source: (IRENA, 2020[9]) (OECD, 2022[2])

In emerging economies such as Mongolia, policy efforts to support low-carbon technology innovation and diffusion must address the reliance on core technology developments from other countries (IEA, 2023[10]). At present, hydrogen innovation (not just related to production, but also, for example, applications in industry and transport) is concentrated within large, technologically advanced firms and across a relatively small number of countries, all of which are high income with the exception of China. For example, as shown in a joint OECD-Joint Research Council of the European Commission report in 2021 of the top 2,000 firms by business research expenditure (BERD), accounting for some 87% of global BERD, the top world's top R&D firms hold 70% of all patents in technologies related to climate change adaptation and mitigation, and 10% of trademarks in the same technologies (compared to 63% of all patents and 6% of trademarks); innovation relating to low-carbon technologies is more regionally and commercially concentrated than in general, and this has implications for the inclusivity of the low-carbon transition in emerging economies (Amoroso et al., 2021[11]). For hydrogen, technological developments are more concentrated. Some 72.1% of all IP5 patents for hydrogen technologies were filed by firms in the top 2,000 worldwide by BERD (Figure 2.2a). Although trademarks, which indicate research commercialisation rather than technological breakthroughs themselves, show less concentration, the share of hydrogen-related trademarks owned by these top firms is still high (27.8%). Notably, this percentage is much higher for hydrogen than for other low-carbon technologies (Figure 2.2b).[1] As shown in Figure 2.2c , the regional revealed technological advantages for hydrogen technologies is much higher in Japan, Korea, and to a lesser extent the EU than the world average.

Figure 2.2. Regional and corporate concentration of selected low-carbon technologies

Share of patents (a) and trademarks (b) owned by the world's top R&D investors in three technology focus areas (2016-18).

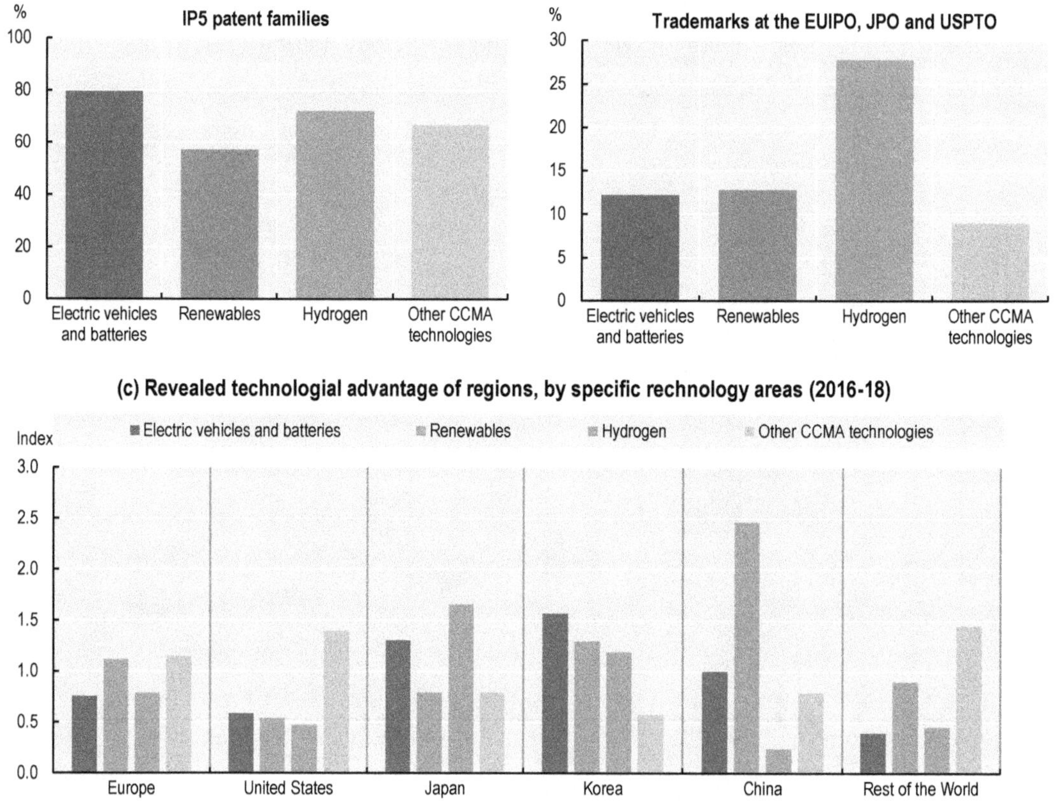

Note: For Figure 2.2a and Figure 2.2b, data reflect the concentration of patenting and trademarking in the selected technology areas among the world's top 2,000 R&D investors, the innovation expenditure of which is equal to 87% of global business R&D expenditures by the private sector and 63% of patent filings across all technologies. For Fig Figure 2.2c, data reflect the revealed technology advantage (RTA) in the same set of technologies at the regional level.
Source : JRC-OECD, COR&DIP© data base v.3., 2021 based on (Amoroso et al., 2021[11])

Any strategy aimed at supporting the development of a renewable hydrogen industry in Mongolia will have to engage with the most effective way of facilitating international technology transfer. The ability of emerging economies to channel innovation taking place abroad, develop it and make it competitive will determine their capacity to become hydrogen producers and exporters. Technology transfer by large, multinational enterprises (MNEs) is an important means of enabling technology diffusion in emerging economies, including for environmental and climate related activities. Debates on the most effective ways of encouraging technology transfer have, in the OECD, often taken place in the context of trade policy discussions, where concerns have emerged around perceptions of 'forcing' technology transfer through local content requirements or other forms of approvals and permitting procedures (Andrenelli, Gourdon and Moisé, 2019[12]). This will also involve the establishing of a framework which enables local firms to benefit from the opportunities that low-carbon technologies present for the local industry. This partly means providing access to lower-cost electrolysers for production, but also ensuring access to innovations in industrial operations. These innovations will be important in enabling the use of hydrogen or its derivatives in Mongolia's hard-to-abate sectors.

The barriers to technology transfer and innovation for achieving Mongolia's low-carbon ambitions are generally accepted in the local science, technology, and innovation community. As part of the country's contribution to the UNFCCC, the National Agency for Meteorology, Hydrology and Environmental Monitoring of Mongolia submitted a position paper on the barriers to the use of internal and external innovation capacities for environmental and climate goals. The paper set out a number of priority policy areas for action that are coherent with OECD analysis and stakeholder consultation. These include identifying sector-specific technology transfer needs, recognising barriers to technology development, transfer and diffusion, exploring ways of participation in bilateral and multilateral mechanisms for technology transfer, and improving the participation of the Mongolian private sector in domestic technology transfer needs (Dagvadorj, n.d.[13]). Similar observations were made during OECD stakeholder consultations in Ulaanbaatar, during which participants noted long-standing but rarely used arrangements within the 2020 Law on Intellectual Property (IPR) for commercialising intellectual property through academic spin-offs and private sector technology transfer. There was a common observation of a lack of awareness within the Mongolian industrial sector regarding the benefits of using low-carbon technologies. This concern was frequently raised both by actors in the STI system as well as in MNEs or their local affiliates. These entities noted the difficulty of integrating their low-carbon activities with the activities of SMEs in their local value chains.

The development of renewable hydrogen requires international co-operation, both in Mongolia and across the OECD. The development of low-carbon technologies is a global effort, requiring significant international co-operation in STI and investment. Within the OECD and G20, such co-operation is typified in large technology development and innovation projects such as the Important Projects of Common European Interest (IPCEIs) in the hydrogen sector and the International Partnership for Hydrogen and Fuel Cells in the Economy (IPHE) (European Commission, 2022; IPHE, 2022). In small, emerging economies such as Mongolia, international co-operation is particularly important. This is due to the anticipated reliance on external technological capacities, financial resources and technical assistance for the development and deployment of domestic low-carbon hydrogen production. Long-standing mechanisms such as the Korea-Israel Industrial R&D Foundation give an example of how countries with shared research priorities and complementary capacities can work bilaterally to accelerate international technology transfer (Israel Innovation Authority, 2024[14]). Given the importance of international innovation and technology co-operation for advanced and developing economies alike, there is a clear rationale for the government to explore how it can better involve its policymakers, technical advisors and the broader STI community in these international discussions.

It is a priority of the leading hydrogen innovator economies to embed their technologies into emerging value chains as part of their own low-carbon transition ambitions. An example of such mechanisms is Japan's Joint Crediting Mechanism, the use of which is being explored by a number of early-stage hydrogen production projects in Mongolia. This mechanism involves a 30% grant (with a ceiling) to cover the purchase of equipment from Japanese technology providers where this enables local emissions reduction (Government of Japan, 2024[15]). More broadly, there is a rich discussion taking place at the global level on fostering co-operation for renewable hydrogen development at a massive scale and ensure that this technology benefits all countries' climate and sustainable growth agendas. To contribute to long-term economic growth in recipient countries, efforts should not only establish renewable hydrogen production sites but also support the development of domestic capabilities to channel this low-carbon technology towards higher value-added products, taking into account local opportunities. In the framework of such co-operation, both the recipient government and its international partners should consider the development of local skills. There is a role here, too, for government, since the ability of Mongolia to benefit from international initiatives to diffuse low-carbon technologies and improve access to the most competitive and locally-suitable innovation will require the government to engage at different levels and from different policy areas at an official bilateral and multilateral level.

Mongolia's existing domestic technology transfer channels can also support the country's renewable hydrogen ambitions. While international co-operation is and will remain important to diffuse low-carbon technologies globally (Box 2.2), domestic channels of knowledge and technology transfer are also important, particularly since such channels can more accurately respond to the particular needs and capacities of local firms. Mongolia already has a number of technology transfer systems in place, notably through the Mongolian Academy of Sciences (the Academy). As discussed in consultations with the OECD in the preparation of this report, the Academy has programmes aimed at supporting academic spin-offs and fostering linkages between industry and academia. These initiatives are designed to facilitate the transfer of knowledge and technologies developed within the domestic public research system to industrial users. These platforms are, however, neither well-known nor well-used, but they nevertheless provide a strong base on which a technology-transfer component of a national hydrogen strategy could be developed. Building on these platforms that already exist, there are a number of things that the Mongolian STI authorities could consider in order to improve the transfer of knowledge and technology between the domestic STI system and the country's industrial sector, particularly its smaller firms. By simply raising awareness of technology transfers that already exist the government could improve participation of firms in existing programmes and accelerate the diffusion of less expensive and more mature technologies which could already be utilised to reduce the emissions-intensity of industrial output. Other options could include integrating technology transfer explicitly into the mandate of publicly funded higher education and research institutions, fostering industry-research partnerships (for example, between large industrial firms and Mongolian research institutions to address local challenges such as efficient water resource management), or supporting the placement of postgraduate students and researchers in industry could be explored. This approach aims to identify opportunities for hydrogen or other low-carbon technology applications into industrial processes.

Box 2.2. Japan's Joint Crediting Mechanism: an example of international financing for technology transfer

The Joint Crediting Mechanism (JCM), established by Japan, is designed to accelerate the dissemination of advanced decarbonisation technologies, products, systems, and infrastructure. Its primary aim is to utilise Japanese technology to achieve significant reductions or removals of greenhouse gas (GHG) emissions, while concurrently promoting sustainable development in partner countries. The emissions avoided through the JCM projects can be used to contribute towards Japan's GHG reduction targets set out in its Nationally Determined Contribution (NDC). The JCM allows projects in participating countries to receive grants to partially cover the cost of investment into the relevant Japanese decarbonisation technology.

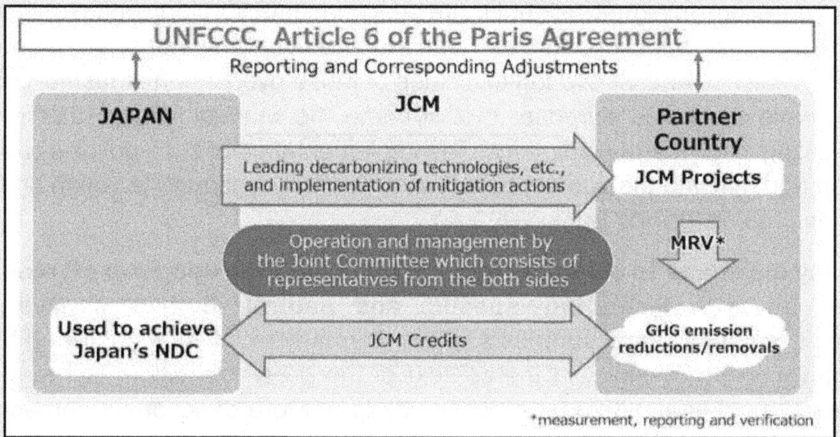

As of January 2024, partnerships with 29 countries demonstrate the JCM's role in promoting global alignment with the goals of the Paris Agreement and the United Nations Framework Convention on Climate Change (UNFCCC). Diverse stakeholders, including Ministry of the Environment of Japan (MOEJ), the Asian Development (ADB) trust fund, and the New Energy and Industrial Technology Development Organisation (NEDO), collectively enhance the JCM's capacity to accelerate the development and adoption of green solutions, like hydrogen-related technologies and infrastructure.

Since Mongolia joined as a partner country on January 8, 2013, nine projects have been developed, expecting to reduce GHG emissions by 75,502 tCO2/year. These projects have focused on renewable energy development and enhancing energy efficiency across various provinces and industries. As Mongolia aims to meet its emission reduction targets and accelerate the shift towards sustainable energy, the JCM has the potential to play a significant role in promoting low-carbon technology adoption, facilitating collaboration, knowledge exchange, and the deployment of innovative solutions, contributing to progress towards a more sustainable future. At least one on-going renewable hydrogen project in Mongolia has utilised the JCM

Source: (Government of Japan, 2024[16])

Stimulating innovation adoption and diffusion through demand-side policies

Domestic innovation in low-carbon technologies and the adoption of low-carbon solutions in domestic industry will depend largely on the signals conveyed by government policies and regulations regarding its commitment to a low-carbon transition. One of the most immediate levers available to governments is to make the cost of emissions higher by using carbon pricing policies and phasing out fossil fuel subsidies. In doing so, the Mongolian authorities can begin to change the relative costs and benefits for competing technologies, while also unlocking new resources to finance climate-related R&D (Cervantes et al., 2023[17]). Such policies force actors to internalise pollution externalities and encourage innovation in and the adoption of more environmentally friendly technologies. Experience from the European Union suggests that the impact of emissions trading schemes on low-carbon innovation can be significant and rapid (Calel and Dechezleprêtre, 2016).

Demand-side interventions are important for renewable hydrogen projects at different levels of development. For firms that might be able to innovate in new uses of renewable hydrogen in existing or new industrial applications, demand-side policies can signal the economic rationale for redirecting R&D activities from existing activities to low-carbon ones, even if the market for the output remains undeveloped. Similarly, for firms that might be able to deploy low-carbon technologies at scale, demand-side policies can support investment decisions in higher-risk but more sustainable technology options, thereby accelerating their use and leading to cost reductions through learning and economies of scale in production. Whether administered through science and innovation ministries or those overseeing industrial development, these policies ultimately have the same purpose: to overcome the market failures that prevent firms from innovating and investing in low-carbon solutions. The implementation of such policies can thereby expedite the green transition and mitigate the substantial off-take risk that continues to weigh on financing costs of renewable hydrogen projects (World Bank, 2023[18]).

Demand-side policy measures to stimulate innovation in (and the adoption of) renewable hydrogen technologies can be both technology-specific and neutral. Both sets have advantages and disadvantages, and it is notable that countries such as Germany with well-established technology-neutral approaches to innovation policy have, in the context of the green transition, increasingly utilised more 'directional' policies (Arnold et al., 2023[19]) (OECD, 2022[20]). More directional approaches to stimulating hydrogen development can take numerous forms. These include government quotas and mandates for the adoption of renewable hydrogen in industrial applications, such as Japan's commitment to having 800,000 fuel cell electric vehicles on the road by 2030, or the EU's target for 42% of hydrogen used in industrial processes to come from renewable fuels of non-biological origin by 2030 (France Strategie, 2018[21]) (European Council, 2023[22])). Additionally, end-use subsidies through, for example, the form of tax credits for the use of renewable hydrogen in hard-to-abate sectors; and public subsidies to cover the capital and operational expenditure of industrial projects for the purchase and operation of renewable hydrogen technology and infrastructure, are also effective mechanisms. Taken together or in isolation, these types of policies can send a clear signal to industrial enterprises and research institutions that there is a developing market for their renewable hydrogen innovation investments, but the extent to which they are actionable by a government will invariably depend on public sector financing constraints and priorities. As discussed in Box 2.3, given the relatively constrained fiscal space available to Mongolian authorities compared to advanced OECD economies, any demand-side policy toolkit should strike a balance. This balance involves conveying the strategic interest of the Mongolian authorities in hydrogen (e.g., by supporting pilot projects, regulating for the use of renewable hydrogen in existing industrial applications) with clear, issue-specific instruments that direct innovation towards outcomes (e.g., lowering emissions from mining) without being overly prescriptive of technology choice.

Box 2.3. Stimulating demand for renewable hydrogen innovation and technology adoption in Mongolia

Demand-side policies can play an important role in de-risking investments into renewable hydrogen innovation and related infrastructure. The challenge for governments, particularly in more fiscally constrained emerging economies, is to strike a balance between sending a clear signal to the private sector that low-carbon technologies are an integral part of economic development policy whilst allowing the private sector to remain free to choose – even if the government has a strategic preference for one technology option over an other – between the options that are most competitive and viable to them. This requires designing a policy toolkit that combines elements of 'directionality' in de-risking instruments (e.g., where they apply to one or a set of technologies) with other levers that are technology neutral (e.g., where they do not specify a technology but rather an outcome, such as lower emissions from a given industry).

Experience from the OECD and IEA demonstrates that demand-side policies aimed at de-risking investment both in specific technologies and their supporting infrastructure are emerging as key components of country strategies for renewable hydrogen development. A non-exhaustive overview of these policies is below.

Carbon Contracts for Difference (CCfD): Governments can spur hydrogen demand by offering CCfD, covering the cost difference between CO2 abatement and a reference price. For example, the UK's CCfD scheme for offshore wind farms has driven investment and innovation, reducing costs over time. Mongolia could replicate this model, guaranteeing stable prices to attract significant investment and enhance hydrogen's competitiveness with fossil fuels.

Public Procurement: Governments can drive renewable hydrogen demand by prioritizing low-emission materials in public procurement. This has proven effective in sectors like construction and transportation, as seen in the US Federal Buy Clean initiative. Mongolia can adopt similar sustainable practices to boost demand for renewable hydrogen in material production processes.

Quotas and Mandates: Governments can stimulate market growth by establishing quotas for low-emission hydrogen consumption. Romania's mandate for industrial users to meet 50% of their hydrogen demand with low-emission sources by 2030 has spurred investment in renewable hydrogen infrastructure. Mongolia could adopt similar policies to drive innovation and decarbonization efforts.

End-Use and CAPEX/OPEX Subsidies: Government subsidies can expand the renewable hydrogen market. For example, the USD 1/kg H2 tax credit in several US states has boosted end-user purchasing power. Mongolia could introduce similar subsidies promoting the adoption of renewable hydrogen technologies. Establishing grant programs through its Innovation Fund could further support market development in industries like mining and manufacturing.

Carbon Taxes: Applying carbon taxes on CO2 emissions in specific sectors, as seen in Chile and Portugal, can incentivize emission reduction and promote the adoption of renewable hydrogen technologies. Mongolia can emulate Chile's approach by implementing a carbon tax on industries exceeding specified thermal power thresholds, thereby encouraging the transition to cleaner energy sources like renewable hydrogen.

Emissions Trading Systems (ETS): Implementing ETS can effectively regulate CO2 emissions and promote the adoption of renewable hydrogen technologies. Following the examples of over 30 ETS worldwide, including those in China, the European Union, and sub-national entities like California and Québec, Mongolia can establish a similar system to allocate emissions allowances and encourage emission reduction efforts across sectors. By including hydrogen production in the ETS, Mongolia can further incentivize the adoption of renewable hydrogen technologies throughout its economy.

Source: Adapted from (IEA, 2023[23])

It is important that such support does not overly determine the availability of other viable technologies and solutions for decarbonisation. The balance between 'directionality' and 'technological neutrality' is at the heart of many ongoing debates around innovation and industrial policy in the context of decarbonisation, where the emphasis is placed on the broader need to compress the innovation cycle for low-carbon technologies and to redirect investment to help scale up promising solutions (OECD, 2023). Technology-neutral policies are also varied, but they often share a rationale of addressing an outcome rather than a particular technology. For example, the use of Carbon Contracts for Difference (CCfD)sees governments covering the difference between the CO2 abatement cost and an average wholesale price to support the de-risking of renewable energy projects. In jurisdictions such as the UK, CCfD schemes are now being applied to hydrogen projects (IEA, 2019[24]). Conversely, effective carbon pricing, whether through emissions trading or other regulatory mechanisms such as emissions standards, can help to lower the competitiveness of fossil-fuel energy sources or carbon-intensive technologies without prescribing a particular technological solution. Additionally, governments can add conditionalities to public procurement to make the purchase of goods and services contingent on specific climate or environmental criteria (IEA, 2023[23]).

There is significant scope for Mongolia to develop demand-side policy signals for low-carbon technology development, for example through green public procurement. Many of the levers used to establish framework conditions for low-carbon investment and market creation in other countries – effective carbon pricing, green quotas, carbon credits – do not yet exist in Mongolia. Using green public procurement as a lever to stimulate demand for low-carbon technologies could work well in Mongolia. The government retains a major stake in many of the most carbon-intensive and environmentally polluting industries, notably in the mining sector, where the government has both a right to a 50% equity stake in strategic mineral deposits and directly controls some of the country's biggest mine operations. A corollary of this public presence is that there is significant scope for the government to directly influence the investment decisions of some of the country's major industrial and infrastructure investors towards more environmentally sustainable and low-carbon outcomes. However, there is currently no clear regulatory or policy push to do so, creating a misalignment between SOE incentives and government climate objectives (De Kleine Feige, 2021[25]). Beyond the existing framework for environmental licensing and permitting, Mongolia could, for example, introduce explicit conditionalities for the use of environmentally sustainable or low-carbon options in public procurement. Several renewable hydrogen developers interviewed for this report are eyeing the mining sector as a client for their pilot projects. Linking renewable hydrogen and mining makes sense on several accounts, as argued in Chapter 1 and other parts of this report: there is a strong rationale for providing public support to establish such linkages as part of a national renewable hydrogen strategy.

The mining and transport sectors may be areas where the government could define a limited set of demand-side interventions to stimulate innovation in and uptake of renewable hydrogen. In conversations with project developers and financial institutions for the preparation of this report, it is clear that a lack of domestic off-take opportunities within Mongolia is a major impediment to de-risking renewable hydrogen projects in the country. While the domestic market in Mongolia for hydrogen is presently small (to the extent that it is used at all in industrial processes, it is generally used as a coolant in thermal power plants, such as in Ulaanbaatar's 663MW Thermal Power Plant 4), the mining and transport sectors are areas where renewable hydrogen could feasibly be used (Table 2.1); encouraging the use of the technology in these areas would have the twin benefit of stimulating demand for hydrogen production in Mongolia whilst also contributing to their decarbonisation in line with the country's NDC. It is notable that the first private sector pilot project to secure an off-take agreement in Mongolia did so within the mining sector, and that it is in the transport sector – specifically in urban mobility – that the Ministry of Energy is focussing its initial efforts to develop domestic renewable hydrogen usage. Given the large SOE footprint in both these sectors, the government could also explore how to use such firms to conduct limited pilot projects while the broader framework conditions for private sector adoption of low-carbon technologies matures.

Table 2.1. Technological Readiness Level (TRL) of hydrogen technologies and their uses in industry and transport

Industry		Transport	
Technology	TRL 2023	Technology	TRL 2023
Hydrogen (low temperature heating)	9	Standard flow rate (60-120 g/s)	9
Electrolytic hydrogen-based produced with variable renewables (ammonia)	8	Hydrogen fuel cell light commercial vehicle	9
Methane pyrolysis (ammonia)	8	Hydrogen fuel cell passenger car	9
Synthetic hydrogen-based fuels in a conventional steam cracker (high value chemicals)	7	Polymer electrolyte membrane hydrogen fuel cell	9
CO2- and electrolytic hydrogen-based produced with variable renewables (methanol)	7	Hydrogen fuel cell urban transit bus	9
Hydrogen (high temperature heating)	7	Methanol-fuelled ship engine	9
Electrolytic hydrogen partially replacing injected coal (blast furnace)	7	Hydrogen fuel cell electric vehicle	8
Based on natural gas with high levels of electrolytic hydrogen blending (DRI)	7	Proton exchange membrane	8
Ammonia (high temperature heating)	5-6	Hydrogen-fuelled truck	7
Based on 100% electrolytic hydrogen (DRI)	6	Molten carbonate	7
Partial use of hydrogen	5	Solid oxide	7
Battery designed for recycling	5	Hydrogen-fuelled light commercial vehicle	6
Chemical absorption - Process gas hydrogen enrichment and CO2 removal for use or storage (blast furnace)	5	Hydrogen-fuelled passenger car	6
Hydrogen for high-temperature heat for ancillary steelmaking processes	5	Hydrogen-fuelled urban transit bus	6
Hydrogen for high-temperature heat for ancillary processes	4	High temperature proton exchange membrane	6
Hydrogen in the Bayer process	4	Methanol fuel cell electric ship	6
Smelting reduction based on hydrogen plasma	4	High flow rate (>120 g/s)	4
Ammonia (low temperature heating)	3		
Using ammonia as reductant (DRI)	2		

Note: Adapted from (IEA, 2019[26]) and (IEA, 2023[27]). The TRL scale is as follows: (1) Initial idea: basic principles have been defined; (2) Application formulated: concept and application of solution have been formulated; (3) Concept needs validation: solution needs to be prototyped and applied; (4) Early prototype: prototype proven in test conditions; (5) Large prototype: components proven in conditions to be deployed; (6) Full prototype at scale: prototype proven at scale in conditions to be deployed; (7) Pre-commercial demonstration: solution working in expected conditions; (8) First-of-a-kind commercial: commercial demonstration, full-scale deployment in final form; (9) Commercial operation in relevant environment: solution is commercially available, needs evolutionary improvement to stay competitive; (10) Integration at scale: solution is commercial but needs further integration efforts; (11) Proof of stability: predictable growth.
Source: (IEA, 2023[27])

Renewable hydrogen has the potential to decarbonise both mine operations and later stages of mineral value chains. As discussed in a 2021 paper by the NewClimate Institute, the mining sector is one potential avenue to stimulate demand for renewable hydrogen in Mongolia, notably through the potential to use fuel cell trucks for mine operations and long-haul transport (Nilsson et al., 2021[28]). Certain existing mine operations remain difficult to decarbonise through electrification (e.g., large vehicles used within the mine, though there are nevertheless efforts to decarbonise these through direct electrification) and they are therefore in need of alternative technology options – of which hydrogen is emerging as one option – to support their decarbonisation. It is notable, however, that the competitiveness of the technology options set out in the NewClimate Institute assume a high-level carbon pricing (USD 100/tCO2), which as discussed in this report does not yet exist in Mongolia. This highlights the need of the government to see the broader framework conditions for the decarbonisation of industry as an integral rather than separate component of any national strategy to develop renewable hydrogen production and usage in Mongolia

There are also opportunities to embed hydrogen-related technologies into future government plans to develop industrial activities in later stages of mineral extraction and processing. At present Mongolia extracts a large quantity of copper (a quantity that will expand following the opening of the underground mine operations at Oyu Tolgoi), but all of the smelting (i.e., the energy intensive but higher value-adding process) takes place in China. It is a priority of the Mongolian government to increase the share of minerals processing within Mongolia, which will necessitate the construction of new infrastructure. For example, the German copper smelter Aurubis is set to become the first to use hydrogen as a substitution for fossil fuels following a EUR 40 million investment to convert its anode furnaces for use hydrogen. While the immediate economics of the Aurubis hydrogen project are influenced by the relatively high level of carbon pricing in the EU – and its future trajectory – it nevertheless is illustrative of a number of considerations for Mongolian policymakers. The first, is that the decision reflects a growing awareness among the minerals sector of the need to decarbonise their entire value chains, and of the importance of doing so both for climatic and competitiveness rationales. The second is that while the ultimate aim of the project is to use only low-carbon hydrogen, the initial phase of the project allows for a flexible input of either gas or hydrogen. The advantage of emerging economies in general is that much of their infrastructures is yet to be built. This which means that options allowing for fuel and energy input flexibility are easier to put in place, although they often mean higher investments.

The development of policies that facilitate a market for low-carbon technologies and fuels would help to foster low-carbon innovation and industries in Mongolia. One theme that runs through much of the OECD's recent work on attracting investment to low-carbon technologies and innovation is the importance of demand-side policies that can de-risk and create incentives for investment in low-carbon projects (Saygin and Lee, 2023[29]) (OECD, 2022[2]) (Cervantes et al., 2023[17]). Such policies can take numerous forms, from reforms to energy feed-in tariffs – as was successfully done by Germany to lower the cost of solar and wind energy – thereby lowering the cost of production inputs to the direct use of public procurement to give preference to low-carbon solutions. Policymakers can also support the development of mechanisms such as previously mentioned CCfDs or providing direct offtake from early-stage projects. It is notable that in Mongolia some of the most ardent proponents of these measures are industrial firms that are seeking a clear framework and policy certainty to invest in solutions for decarbonisation. However, the government has not meaningfully defined and adopted the types of demand-side policies and regulatory reforms that would give incumbent and future investors certainty. The experience of the introduction of Feed-in Tariffs and later disputes and concerns over their affordability and flexibility perhaps highlight the need for the government to ensure that any publicly-backed demand-side intervention is financially sustainable and that there is a vision for the competitiveness of low-carbon technology investments that is not dependent on long-term subsidy. The introduction and linkage of revenue-raising mechanisms, such as a modest carbon price floor, could be used to support government interventions aimed at stimulating the use of low-carbon technologies such as hydrogen in industry and other sectors.

Managing the interaction of the various policy areas necessary to redirect public and private sector investment towards low-carbon outcomes may require new institutional approaches and capacities. Recent work by the OECD has demonstrated, for example, that the success of innovation and industrial policies in the context of the low-carbon transition depends on new forms of cross-sectoral policy coordination; such coordination has always been positive for policymaking, but in achieving decarbonisation it is increasingly essential (Arnold et al., 2023[19]). In practice, this is due to the need to carefully balance a range of policy instruments across – though not exclusively – fiscal policy, investment preferences, environmental regulation, industrial incentives, skills and education, and infrastructure investment. In Mongolia, and in the context of the country's hydrogen ambitions, this will require raising the knowledge of public servants and decision-makers on the technological, infrastructure and economic implications of hydrogen development and varying scales, and improving the coordination across the various areas of policy-making – from heavy industry and energy to fiscal policy and STI – that are necessary to bring together in order to give the country's nascent hydrogen industry the best chance of success This is a challenge for emerging economies such as Mongolia as well as in the advanced economies of the OECD, and in practice it may mean changing or expanding the scope of sector-specific policy committees or agencies to better allow for the integration of experience from adjacent policy areas in policymaking and implementation.

Aligning Mongolia's innovation and industrial policy framework on low-carbon targets

The policy and regulatory framework for innovation in Mongolia

Government ambitions for innovation in Mongolia are set out a number of strategic documents, with some of these having explicit low-carbon development objectives. There are two recent strategic documents that are of particular relevance for framing public policy interventions to support the low-carbon transition in Mongolia. The first of these is the 'Vision 2050 Long-Term Development Policy of Mongolia', which contains a set of explicit objectives for developing the country's STI system, ensuring that STI contributes to the sustainable economic development of Mongolia. These objectives are split over three stages, with the first (2021-2030) focussing on: creating an enabling environment between public research and science and the private sector, ensuring a mixed source of financing for this relationship; expanding STI infrastructure; reforming incentives for researchers and scientists; and increasing their participation in international research collaboration. The second stage (2031-2040) aims to increase funding and resources for research in a number of frontier technologies – e.g., artificial intelligence and technologies to support advanced manufacturing – and to integrate international standards in these areas into the national quality infrastructure framework.

These ambitious commitments were further developed in the government's 'New Recovery Policy 2021' (NRP). This document was drafted to steer the socio-economic development in the wake of the global COVID-19 pandemic (Government of Mongolia, 2023[30]). As part of the NRP, the government integrated a commitment to developing technological and innovation capacities into its industrial revival pillar, linking these efforts to the high-level government ambition of creating more value in the country's agricultural sector. Nevertheless, throughout the sub-indicators and actions of other priority areas in the industrial components of the NRP there are a number of references to the importance of science- and knowledge-based approaches. Notably, the document explicitly references the development of pilot renewable hydrogen projects in the South Gobi region as part of the government's ambitions to improve the resilience and reduce the carbon-intensity of the energy system. This latter point is important in so far as it highlights the polyvalence and complementary rationales for hydrogen development in Mongolia; as a means of industrial decarbonisation and as a potential way to integrate more renewable energy into other applications (such as, for example, providing dispatchable power from VRE sources in electric vehicle charging infrastructure). There has also been a partial institutionalisation of the NRP through the New Recovery Policy Accelerator, whose mandate is to provide policy support to projects that can advance the objectives of the NRP; at present it has no low-carbon projects, but OECD consultations with stakeholders in Ulaanbaatar suggested that the institution may play a role in supporting pilot and demonstration projects in renewable hydrogen production.

These recent developments build on a general increase in attention paid by Mongolian policymakers to the contribution of science, technology, and innovation (STI) to socio-economic outcomes. The government has received technical assistance from the Asian Development Bank (ADB) to improve the policy and regulatory framework for STI and its contribution to economic development.

Concretely, the NRP follows from the development of the Action Program of the Government of Mongolia for 2016-2020, in which the government set out its vision for improving the legal and policy frameworks for innovation, innovation financing, research and data infrastructure. This vision was reiterated in the subsequent 'Priority Areas for Innovation (2020-2025)' policy programme, where the government emphasised its ambition for innovation- and knowledge-based growth.

The government has articulated a linkage between the need to develop renewable energy capacities and investment and innovation policies. For example, in the aforementioned 'Priority Areas for Innovation (2020-2025)' policy programme, the government sets out the need to attract investment and develop innovation capacities in order to support both the production of renewable energy as well as the diffusion of energy efficiency technologies to minimise heat loss from. Importantly, the same document also emphasises renewable energy storage to increase reliability in the energy system, with the development of battery storage solutions – pilot projects for which are under development with the ADB – viewed by many stakeholders interviewed for this report as a crucial step in improving the integration of VREs into the Mongolian energy system (Government of Mongolia, n.d.[31]).

There is a fairly well-developed legal framework for science and innovation activities in Mongolia, with the government undertaking legal reforms to improve the framework for intellectual property development and transfer. Significant pieces of legislation include the 2006 Law on Science and Technology and the 2012 Law on Innovation, which is in the process of being amended. The Law on Science and Technology entails 6 different chapters covering a variety of issues from funding for science and technology activities to provisions ruling the social security of the researcher. It also establishes the mandate of the Science and Technology Council of the Ministry over the sector. According to the law the Council is in charge of identifying scientific and technological priorities, implementing measures to strengthen scientific and industrial alliances and cooperating with universities and research institutes, among others.

A major development has been the adoption of the 2020 Law on Intellectual Property (IPR), which superseded an earlier 2006 law. As noted by the ADB in 2017, Mongolia has faced significant challenges with IPR-based knowledge and technology transfer, notably around the methodology for IPR valuation and a lack of quality assurance in the granting of IPR. As discussed below, given the significant role that publicly-funded research plays in the Mongolian innovation system, these shortcomings have created challenges around IPR ownership, posing issues for the commercialisation of research through arrangements such as academic spin-offs and joint ventures. The 2020 IPR law may therefore prove to be an important part of the regulatory framework for low-carbon technology development in Mongolia, since it seeks to address a number of shortcomings in the previous IPR framework and thereby the conditions for technology transfer (Government of Mongolia, 2020[32]).

The institutional arrangement for the innovation system is relatively fragmented. The Ministry of Education and Science (MES) is responsible for the overall design and implementation of STI policy in the country, and directly supervises the majority of the 59 public research institutions and the 21 research-based universities. Of Mongolia's public and private research institutions, around one third (20) are supervised by other line ministries. Additionally,10 are under the supervision of the Mongolian Academy of Sciences, which is affiliated with the MES. While such complex institutional arrangements are not uncommon in innovation systems, they can hinder the formulation of policies aimed at directing research and innovation resources toward complex, systemic challenges. This is particularly evident in Mongolia, where limited R&D resources are spread across a wide range of uncoordinated institutions, leading to a misalignment of these resources with local industrial needs and government policy objectives.

The Mongolian research base and its alignment with industrial needs

Mongolia's innovation system is small in relative and absolute terms, which limits the extent that the development and deployment of low-carbon technologies in the country can draw on domestic

resources. Mongolia's research base – meaning the strength and competencies of the institutions and their personnel dedicated to research in the public and private sectors – is an integral part of any innovation system. This base can provide expertise, capabilities, can raise awareness, and can partner with firms in innovation activities; this can be crucial for SMEs which lack internal R&D capacities. In the context of industrial decarbonisation, where industrial firms will increasingly require new forms of expertise and cross-disciplinary collaboration (e.g., traditional production equipment in cement manufacturing with new low-carbon fuels or carbon capture facilities), the strength of this base and the ease with which industrial SMEs can make use of it is likely to grow in importance. As noted in the Vision 2050 document, the government has an ambition to strengthen the linkages between the private sector and public research, but at present it is not clear if there are any explicit policies – e.g., support for co-financed research between public and private actors, academic placements in industry, SOE contracting of R&D from domestic research institutions – that have been implemented to this end.

Despite government commitments to raise levels of innovation expenditure, levels of R&D expenditure in Mongolia remain also low and have fallen in relative terms over time. While there is limited data available on firm-level R&D expenditure in Mongolia or disaggregated data on R&D by area of activity, there does exist high-level gross R&D expenditure (GERD) data, which is collected under SDG 9.5.1. In the most recent year for which data are available, GERD in Mongolia was equal to 0.1% GDP, similar to levels in Kazakhstan and Uzbekistan (0.12 and 0.13% GDP respectively), but far below the OECD average of 2.57% GDP (Figure 2.3). Perhaps of greater concern is that since over the past two decades GERD in Mongolia as a share of GDP has fallen (from 0.19% GDP in 2000). The general trend across the OECD has been for intramural expenditure on R&D to have increased as a share of GDP.

Figure 2.3. (SDG 9.5.1) Gross R&D expenditure (%GDP), 2000 and 2022

OECD members and SIPA Central Asia countries (Kazakhstan, Mongolia and Uzbekistan)

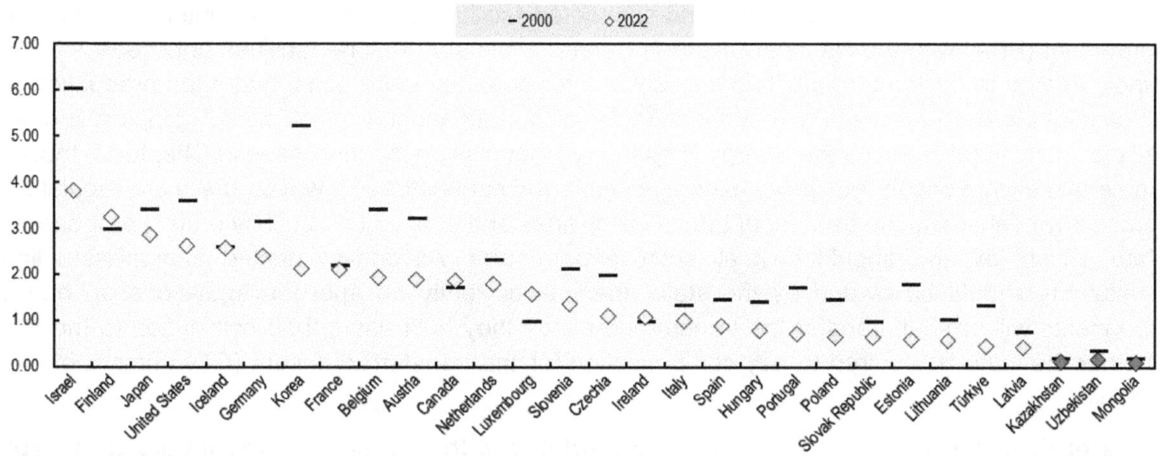

Note: OECD countries for which 2022 data are not available have been excluded.
Source: (OECD, 2024[33]) (World Bank, 2023[34])

The contribution of Mongolian innovation to economic and low-carbon development is likely held back by a limited supply of research capacity. The share of fulltime researchers (an indicator of the magnitude if not the quality of the research capacities in a given economy) in Mongolia is low in international comparison, below Kazakhstan and Uzbekistan (626 and 525 fulltime researchers per million population compared to 332 per million in Mongolia), with all three significantly below the OECD average of 4079 researchers per million population; a similar gap exists between Mongolia and other large mining economies, such as Canada and Australia (4594 and 5076 per million population respectively) (World Bank, 2023[34]). In absolute terms, Mongolia had 2515 fulltime research personnel across the public and private sector in 2015, a decline from 3102 in 1995. Data on the allocation of research staff within Mongolian industries, as well as on industry-science linkages (academic-industry placements, joint PhD supervision in key industries, etc.) are not available.

Operational environment for low-carbon technology projects

The operational environment for firms involved in renewable hydrogen projects – i.e., the regulatory, licensing and permitting requirements governing such projects– is governed by a range of laws and regulations, each addressing different aspects of project development. The Land Law (1994, last amended in 2002), as the primary legislation governing land tenure and licensing, plays a crucial role since most land in Mongolia is state-owned and land rights are essential for hydrogen project development (Government of Mongolia, 1994[35]). Experience from other emerging economies with ambitions to become major hydrogen producers has shown that simplifying access to land is one of the major interventions governments may take to support the de-risking of hydrogen projects (IEA, 2023[36]). At present, there is no specific regime for hydrogen projects in the legislation. Thus, such projects would be governed by general land allocation rules and procedures. Upon obtaining land permits, fees must be paid as outlined in the Land Fees Law (Goverment of Mongolia, 1994[37]). Additionally, the Minerals Law and the Petroleum Law grant mineral license holders and parties certain powers to obtain land surface rights; in the cast of permitting for renewable energy projects. However, there may be some potential inconsistencies between the Minerals Law and the Renewable Energy Law, which requires that land permits must have been already obtained by the applicants before applying for the needed licences and permission (Government of Mongolia, 2006[38]) (Government of Mongolia, 2014[39]).

The Renewable Energy Law categorises renewable energy plants into two types: those connected to the transmission grid and stand-alone facilities. This classification affects the types of licenses and agreements developers must secure. Grid-connected facilities are required to enter a Power Purchase Agreement (PPA) with the National Dispatch Centre, while stand-alone facilities only need a PPA if they supply energy to customers; this has implications for potential large-scale hydrogen production projects which might wish to secure a grid connection to sell surplus electricity. All energy projects are bound to sell electricity at rates set by the Energy Regulatory Commission, as discussed in Chapter 3, though there are certain exceptions for electricity exports without grid connection. However, there are regulatory gaps, particularly concerning the transfer of transmission lines and substations built by project proponents to the State, which, by law, should own all such infrastructure. All grid-connected transmission lines and substations should be owned by the state, there is currently no specific regulations on how project proponents will or shall transfer the transmission lines they built using their own funds to the state and whether the state is required to pay compensation for the transferred assets (Government of Mongolia, 2015[40]).

While project developers are relatively comfortable with the existing regulatory framework, the nature of supporting frontier technologies such as renewable hydrogen may require a degree of experimentation and agility. If Mongolia wishes to position itself as a first mover in renewable hydrogen production, the government must find a balance between establishing a robust regulatory framework for the safe production and use of renewable hydrogen and maintaining a degree of regulatory agility, based on real risks and up-to-date evidence, to support innovation. In practice, this means ensuring that the policymakers and institutions in Mongolia that are responsible for the development of the regulatory framework for low-carbon technologies have the capacities and mandate to adapt their regulatory approach as projects develop and new evidence becomes available (OECD, 2023[41]).

The OECD Recommendation on Agile Regulatory Governance to Harness Innovation may be instructive for Mongolian policymakers. The Recommendation contains insights that could inform reforms in Mongolia to improve the suitability of the regulatory frameworks for the type of innovation and technology adoption necessary for success in the low-carbon transition. This is particularly relevant when innovation occurs at the intersection of technologies or industries previously regulated separately or under different assumptions (for example, the use of hydrogen in pre-existing industrial applications as opposed to its use in applications where the rationale is emissions reduction) (OECD, 2021[42]). As noted in the Recommendation, there is a recognition across the OECD of the need to adjust regulatory management tools to ensure that regulation is fit for future purposes. This includes adopting more iterative and flexible regulatory assessment cycles and improving co-operation between different sectoral regulatory agencies. The underlying rationale, which centres on accelerating innovation in green and digital technologies, is as relevant to Mongolia and non-member OECD economies as it is to advanced OECD economies.

The development and oversight of the regulations and licenses governing hydrogen, as for other chemicals and fuels, is overseen by the National Energy Centre (NEC) of Mongolia, under the Ministry of Energy. The NEC has significant experience with fuel-related regulation and inspection but will likely require capacity-building to ensure that it can support the development of a wider use and production of renewable hydrogen in the country. Projects involving hydrogen, classified in Mongolian legislation as a hazardous chemical, falls under the Toxic and Hazardous Chemical Law, necessitate a special license for its production, storage, trade, and use (Government of Mongolia, 2006[43]). As across the OECD, Mongolia will likely need to update these provisions to allow for the production and use of hydrogen as a fuel and in new industrial processes in a way that promotes a rational use of natural resources, sets clear limits on pollution, and mitigates adverse environmental impacts more broadly.

Environmental regulatory considerations are also important for any potential hydrogen project. Environmental Impact Assessments (EIAs) are mandatory for projects involving toxic and hazardous chemicals (as hydrogen is currently designated), as well as for solar and wind farms. At present, there is no legal requirement to conduct a strategic environmental assessment (SEAs), that is, an assessment of the potential broader, systemic impact of a project beyond the immediate project site. Given the interconnections between land, water, and energy use that go beyond the immediate project site for hydrogen production (e.g., impact on downstream river basin if drawing water from Kherlen river), or the unclear impact of hydrogen production at scale on groundwater reserves, it may be advisable for the government to apply a SEA framework at the policy or strategy level. As discussed in Chapter 4, given the social sensitivities in Mongolia around industrial water use, this may also politically de-risk any large investment project in Mongolia's hydrogen sector.

Recent developments in Mongolian innovation relevant to the development of low-carbon technology capacities

Although Mongolia faces challenges such as a fragmented institutional environment, limited R&D investment and a lack of policy signals and guidance for low-carbon technology innovation, there are nevertheless a number of STI developments of relevance to renewable hydrogen development in the country. In terms of publicly financed research, Mongolia has a long-established hydrogen and other low-carbon technology initiatives at the Academy of Sciences, where there are ongoing projects on fuel-cell technologies and circular economy solutions for electric vehicle batteries. This work is conducted by faculty and postgraduate research students across a number of technology-specific laboratories within an institution with spinoff and transfer networks; participants in this work have been involved in efforts under the Hydrogen Council of Mongolia to establish a coherent and unified approach to hydrogen development int the country. The hydrogen-related work was initiated at the government's request in 2010, but there has been no significant change to the scope of this request since then. The Academy has provided ad hoc advice to the government on hydrogen policy issues. However, there has been very little commercialisation of the hydrogen work undertaken at the Academy, and engagement with industry on their hydrogen-related projects is limited.

International firms and large, tradeable industrial firms are undertaking research and innovation to support the decarbonisation of their value chains. A number of observations emerged from industrial stakeholder consultations that are relevant to Mongolian policymakers considering how best to support the emergence of a renewable hydrogen industry in the country. Large mining enterprises, both domestic and foreign-owned, are undertaking significant R&D and infrastructure investments to ensure that their production is sustainable and aligns with evolving international emission intensity regulations. Hydrogen is one of the technologies being explored by such firms – one of the pilot hydrogen projects has concluded a Memorandum of Understanding (MoU) with one such firm to demonstrate viability (Box 2.4) – but it is not the only one. Crucially, firms in the mining sector, which present a viable use-case for renewable hydrogen under certain market conditions, are also actively exploring electrification of their heavy vehicles and processing operations. This is true both of firms with Mongolian and foreign ownership and is indicative of the risks involved with overly prescriptive innovation and technology policies, in that the private sector may choose an alternative approach to managing their decarbonisation. Furthermore, industrial firms have emphasised the need for stronger policy signals from the government on effective carbon pricing and grid decarbonisation. Such signals would enable them to rationalise the large investments necessary to adopt low-carbon technology and fuel options such as renewable hydrogen. Despite an absence of clear policy signals for decarbonisation within Mongolia, certain firms – foreign and domestic – are nevertheless redirecting investment to low-carbon options in anticipation of the impact of external regulatory developments on the competitiveness of their production.

Box 2.4. An example pilot project of renewable hydrogen in Mongolia: Gobi H2

SunSteppe Renewable Energy ('SRE'), a joint venture between Petro Matad Limited and partners, is developing the first hydrogen demonstration project in the South Gobi region of Umnugobi aimag, Mongolia. The project aims to achieve financial close by late 2024, with construction scheduled to begin in spring 2025, and commissioning proposed for Q1 2026. Strategically located near key coal and copper mining operations, the project will provide green hydrogen to serve Oyu Tolgoi, one of the world's largest copper and gold deposits, and associated camps.

A Memorandum of Understanding has been established with Oyu Tolgoi and the Japanese Overseas Environmental Cooperation Centre ('OECC') to develop the project and supply green hydrogen for heating facilities at the mine. This partnership with OECC has already secured some Japanese Government grant funding, with additional grant applications expected in the future. This funding could cover up to 40% of the capital expenditure required for the project, significantly enhancing its financial viability.

The financial structure of this project, comprising both debt and equity instruments, is in the final stages of being finalised. Equity funding is led by Wolfson Energy in partnership with Petro Matad, with potential third-party participation under consideration. Discussions with potential lenders for debt financing are ongoing, alongside exploration of financing avenues linked to technology providers. This collaboration highlights the commitment of both public and private sectors to promoting renewable energy in Mongolia.

Source: Sunsteppe/Petro Matad Limited case study (Annex A)

A significant driver of hydrogen related research in Mongolia has come through the Industrialisation Policy Committee, which was established in early 2023 to support the government with its industrialisation and diversification objectives. The clearest evidence of the impact institutional development on hydrogen innovation in Mongolia has been the direction of work undertaken by a working group on hydrogen development that was set up under this committee. The vice chair of the working group is a senior industrial figure in the country and CEO of one of Mongolia's largest industrial and mining firms, MAK LLC (MAK). At the initiative of the working party, MAK is currently undertaking one of the largest hydrogen-related projects in Mongolia (Annex A). The MAK project is illustrative of a number of considerations for policymakers in Mongolia interested in supporting hydrogen development. The first is that, while concentrated in a small number of firms, industrial firms in the private sector possesses significant innovation capacities. The second concerns the challenges of demand-creation and market maturity, with the firm using its own internal purchasing power to create a small market for the output of its R&D. The third is that mixed signals on coal phase-out may lead to innovation activities whose output is divergent with international best practices and standards for low-carbon technologies (See Table 3.1 in Chapter 3).

'Blue' hydrogen production in early stages of renewable hydrogen development is a possible way of creating a market for domestic hydrogen, but only under strict conditions. For example, if the MAK project proves to be economically and technologically viable, then it could be an important, albeit time-bound and intermediate, step in Mongolia's low-carbon transition. In fact, OECD work has discussed the use of CCUS and fossil-fuel produced hydrogen as a possible approach for a gradual building of hydrogen capacities in emerging markets. Nevertheless, policy support for blue hydrogen projects risks mixed signals regarding the government's commitment to coal phase-out and creating an investment environment that privileges low-carbon technological development over brown alternatives (OECD, 2022[2]). As noted in consultations with the OECD, firms may question the rationale to develop renewable hydrogen projects if the small current demand for hydrogen could be met with blue projects. There are therefore risks involved in pursuing blue hydrogen as an intermediary step, which policymakers should take into consideration when providing public support. This also applies to the competitiveness of hydrogen projects in Mongolia on external markets. For example, depending on the standards that become the benchmark for international hydrogen trade, there is a risk that investment in blue hydrogen projects results in stranded assets, such as additional coal-related infrastructure. Similarly, if policy decisions are taken to continue coal exploitation under an assumption that coal assets can be used for hydrogen production, this too might lead to a misallocation of finance that could have been used for the development of other low-carbon infrastructure and technologies. Fundamentally, policymakers and industrial firms in Mongolia must balance the need to support local industrial development leveraging existing assets and capacities with the need to align production with standards and certifications that will ensure that Mongolian production remains viable and competitive in external markets.

References

Amoroso, S. et al. (2021), *World Corporate Top R&D Investors: Paving the way for climate neutrality, A joint JRC and OECD report*, Publications Office of the European Union, Luxembourg, https://web-archive.oecd.org/2021-11-22/617125-world-corporate-top-rd-investors-paving-the-way-for-climate-neutrality.pdf. [11]

Andrenelli, A., J. Gourdon and E. Moisé (2019), *International Technology Transfer Policies*, OECD Trade Policy Papers, OECD, Paris, https://www.oecd-ilibrary.org/docserver/7103eabf-en.pdf?expires=1705592690&id=id&accname=ocid84004878&checksum=B33304F31E9A25F160247B1F57800ADD. [12]

Arnold, E. et al. (2023), *Navigating green and digital transitions: Five imperatives for effective STI policy*, OECD Science, Technology and Industry Policy Papers, OECD, Paris, https://www.oecd-ilibrary.org/science-and-technology/navigating-green-and-digital-transitions_dffb0747-en. [19]

ASDIT (2023), *ASDIT*, https://www.asdit.cl/en/inicio-english/. [8]

Cammeraat, E., A. Dechezlepretre and G. Lalanne (2022), *Innovation and Industrial Policies for Green Hydrogen*, OECD Science, Technology and Industry Policy Papers, OECD, Paris, https://www.oecd-ilibrary.org/docserver/f0bb5d8c-en.pdf?expires=1692286715&id=id&accname=guest&checksum=767FA1D7985F43131BCEA25FFAC2755B. [1]

Cammeraat, E., A. Dechezleprêtre and G. Lalanne (2022), "Innovation and industrial policies for green hydrogen", *OECD Science, Technology and Industry Policy Papers*, No. 125, OECD Publishing, Paris, https://doi.org/10.1787/f0bb5d8c-en. [4]

Cervantes, M. et al. (2023), *Driving low-carbon innovations for climate neutrality*, OECD Science, Technology and Industry Policy Papers, No. No. 143, OECD, Paris, https://www.oecd-ilibrary.org/docserver/8e6ae16b-en.pdf?expires=1705506379&id=id&accname=ocid84004878&checksum=C36BE1F776B1CEF2C7CE1B1EB595E625. [17]

Climate Fund Managers (2022), *Launch Of The New 'SDG Namibia One Fund' At Cop27 Targets The Development Of A Green Hydrogen Economy In Namibia*, https://climatefundmanagers.com/2022/11/09/launch-of-the-new-sdg-namibia-one-fund-at-cop27-targets-the-development-of-a-green-hydrogen-economy-in-namibia/. [69]

CMS (2021), *HYDROGEN LAW, REGULATIONS & STRATEGY IN CHILE*, https://cms.law/en/int/expert-guides/cms-expert-guide-to-hydrogen/chile. [45]

Cordonnier, J. and D. Saygin (2023), *Financing solutions to foster industrial decarbonisation in emerging and developing economies*, OECD Environmnt Working Papers, OECD, Paris, https://www.oecd-ilibrary.org/docserver/24a155ab-en.pdf?expires=1705514321&id=id&accname=ocid84004878&checksum=8B304BE24018556EA4E6F4666DCA9A93. [3]

Criscuolo, C. et al. (2023), *Quanitfying industrial strategies across nine OECD countries*, OECD Science, Technology and Industry Policy Papers, No. No. 150, OECD, Paris, https://www.oecd-ilibrary.org/science-and-technology/quantifying-industrial-strategies-across-nine-oecd-countries_5f2dcc8e-en. [77]

Dagvadorj, D. (n.d.), *Technology Transfer Needs and Barriers in Mongolia*, National Agency for Meteorology, Hydrology and Environment Monitoring of Mongolia, Ulaanbaatar, https://unfccc.int/sites/default/files/moncp.pdf. [13]

De Kleine Feige, A. (2021), *State-Owned Enterprises and Climate Action*, World Bank Group, DC, https://documents1.worldbank.org/curated/en/886671636136930215/pdf/State-Owned-Enterprises-and-Climate-Action.pdf. [25]

Diario Fianciero (2023), *Los detalles de la readjudicacion del Instituto de Tecnologias Limpias de Corfo al consorcio Asdit*, https://www.df.cl/df-lab/innovacion-y-startups/los-detalles-de-la-readjudicacion-del-instituto-de-tecnologias-limpias. [53]

Environmental Investment Fund (2023), *THE GOVERNMENT OF NAMIBIA AND DUTCH PARTNERS INK MEMORANDUM OF UNDERSTANDING*, https://www.eif.org.na/post/sdg-namibia-one-press-release-20-june-2023. [70]

European Commission (2023), *Delegated regulation for a minimum threshold for GHG savings of recycled carbon fuels and annex*, European Commission, https://energy.ec.europa.eu/publications/delegated-regulation-minimum-threshold-ghg-savings-recycled-carbon-fuels-and-annex_en. [49]

European Commission (2023), *Delegated regulation on Union methodology for RFNBOs*, European Commission, https://energy.ec.europa.eu/publications/delegated-regulation-union-methodology-rfnbos_en. [48]

European Council (2023), *Renewable energy: Council adopts new rules*, European Parliament, https://www.consilium.europa.eu/en/press/press-releases/2023/10/09/renewable-energy-council-adopts-new-rules/. [22]

European Parliament (2023), *EU rules for renewable hydrogen*, European Parliament, https://www.europarl.europa.eu/RegData/etudes/BRIE/2023/747085/EPRS_BRI(2023)747085_EN.pdf. [50]

France Strategie (2018), *Overview of government policies to promote ultra-low emission vehicles*, France Strategie, https://www.strategie.gouv.fr/sites/strategie.gouv.fr/files/atoms/files/note_de_synthese_-_vehicules_electriques_-_gb_0.pdf. [21]

Goverment of Mongolia (2007), *Renewable Energy Law*, Goverment of Mongolia. [61]

Goverment of Mongolia (1994), *Land Law*, Government of Mongolia. [37]

Government of Canada (2022), *Minister Wilkinson Announces up to $800 Million in Project Funding to Advance Canada's Clean Fuels Sector*, https://www.canada.ca/en/natural-resources-canada/news/2022/11/minister-wilkinson-announces-up-to-800-million-in-project-funding-to-advance-canadas-clean-fuels-sector.html. [73]

Government of Canada (2020), *Canada Hydrogen Strategy*, Government of Canada, https://natural-resources.canada.ca/climate-change-adapting-impacts-and-reducing-emissions/canadas-green-future/the-hydrogen-strategy/23080. [44]

Government of Canada (n.d.), *Clean Fuels Fund*, https://natural-resources.canada.ca/climate-change/canadas-green-future/clean-fuels-fund/23734. [72]

Government of Chile (2023), *Gobierno presenta Fondo por US$ 1.000 millones para el desarrollo del Hidrógeno Verde en Chile*, Government of Chile, https://energia.gob.cl/noticias/nacional/gobierno-presenta-fondo-por-us-1000-millones-para-el-desarrollo-del-hidrogeno-verde-en-chile. [63]

Government of Chile (2023), *Government presents Strategic Committee for Green Hydrogen Action Plan and measures to promote the development of this industry*, https://www.hacienda.cl/noticias-y-eventos/noticias/gobierno-presenta-comite-estrategico-para-plan-de-accion-de-hidrogeno-verde-y. [74]

Government of Chile (2022), *Government launches website for citizens to participate in the development of the green hydrogen industry*, Government of Chile, https://www.gob.cl/en/news/government-launches-website-citizens-participate-development-green-hydrogen-industry/. [57]

Government of Chile (2021), *Chile's Green Hydrogen Strategy and investment opportunities*, https://energia.gob.cl/sites/default/files/documentos/green_h2_strategy_chile.pdf. [7]

Government of Chile (2020), *Chil Green Hydrogen National Strategy*, Government of Chile, https://energia.gob.cl/sites/default/files/national_green_hydrogen_strategy_-_chile.pdf. [47]

Government of Japan (2024), *Basic Concept of the Joint Crediting Mechanism*, Government of Japan, Tokyo, https://www.jcm.go.jp/about. [15]

Government of Japan (2024), *Joint Crediting Mechanism*, Government of Japan, https://www.jcm.go.jp/about. [16]

Government of Mongolia (2023), *New Recovery Policy*, Government of Mongolia, https://www.nrpa.gov.mn/en/new-recovery-policy. [30]

Government of Mongolia (2020), *Law on Intellectual Property*, Government of Mongolia. [32]

Government of Mongolia (2015), *Energy Law*, Government of Mongolia. [40]

Government of Mongolia (2015), *The Law on Energy Saving*, Government of Mongolia. [66]

Government of Mongolia (2014), *Law of Mongolia on Petroleum*, Goverment of Mongolia. [39]

Government of Mongolia (2013), *Investment Law*, Government of Mongolia. [67]

Government of Mongolia (2012), *Exemption from Customs and Value Added Tax*, Government of Mongolia. [64]

Government of Mongolia (2008), *On Customs Tariffs and Customs Taxes*, Government of Mongolia. [65]

Government of Mongolia (2006), *LAW ON SCIENCE AND TECHNOLOGY*, Goverment of Mongolia. [60]

Government of Mongolia (2006), *Minerals Law*, Govermnet of Mongolia. [38]

Government of Mongolia (2006), *Toxic and Hazardous Chemicals Law*, Government of Mongolia. [43]

Government of Mongolia (1994), *Land Law*, Government of Mongolia, https://faolex.fao.org/docs/pdf/mon4919E.pdf. [35]

Government of Mongolia (n.d.), *Priority Areas for Innovation (2020-2025)*, Government of Mongolia. [31]

Government of Monoglia (2001), *Environmental Impact Assestment*, Government of Monoglia. [62]

Government of Morocco (2021), *FEUILLE DE ROUTE HYDROGÈNE VERT*, https://www.mem.gov.ma/Lists/Lst_rapports/Attachments/36/Feuille%20de%20route%20de%20hydrog%C3%A8ne%20vert.pdf. [5]

Government of Namibia (2022), *Namibia Hydrogen Strategy*, Governmen of Namibia, https://www.ensafrica.com/uploads/newsarticles/0_namibia-gh2-strategy-rev2.pdf. [6]

Green Hydrogen Organisation (n.d.), *Country Portal - Chile*, https://gh2.org/countries/chile#:~:text=Six%20projects%20by%202025,subsidies%20totalling%20US%2450%20million. [71]

Green Hydrogen Organization (n.d.), *Chile's green hydrogen vision*, https://gh2.org/countries/chile. [46]

Green Hydrogen Organization (n.d.), *The Africa Green Hydrogen Alliance (AGHA)*, https://gh2.org/africa-green-hydrogen-alliance-agha. [51]

Hydrogen Europe (2023), *Namibian govt prepares for green H2 legislation*, https://hydrogeneurope.eu/namibian-govt-prepares-for-green-h2-legislation/. [81]

IEA (2024), *Renewables 2023*, IEA, Paris, https://www.iea.org/reports/renewables-2023. [55]

IEA (2023), *ETP Clean Energy Technology Guide*, IEA, Paris, https://www.iea.org/data-and-statistics/data-tools/etp-clean-energy-technology-guide. [27]

IEA (2023), *Global Hydrogen Review 2023*, IEA, Paris, https://www.iea.org/reports/global-hydrogen-review-2023. [23]

IEA (2023), *Hydrogen patents for a clean energy future: A global trend analysis of innovation along hydrogen value chains*, IEA, https://iea.blob.core.windows.net/assets/1b7ab289-ecbc-4ec2-a238-f7d4f022d60f/Hydrogenpatentsforacleanenergyfuture.pdf. [10]

IEA (2023), *Renewable Hydrogen from Oman*, IEA, https://www.iea.org/reports/renewable-hydrogen-from-oman. [36]

IEA (2023), *Towards hydrogen definitions based on their emissions intensity*, IEA, https://www.iea.org/reports/towards-hydrogen-definitions-based-on-their-emissions-intensity. [59]

IEA (2019), *Contract for Difference (CfD)*, IEA, Paris, https://www.iea.org/policies/5731-contract-for-difference-cfd. [24]

IEA (2019), *Innovation Gaps*, IEA, Paris, https://www.iea.org/reports/innovation-gaps. [26]

IRENA (2020), *Green hydrogen cost reduction*, https://www.irena.org/publications/2020/Dec/Green-hydrogen-cost-reduction. [9]

Israel Innovation Authority (2024), *The Korea- Israel Industrial R&D Foundation (KORIL-RDF)*, Israel Innovation Authority, https://innovationisrael.org.il/en/foundation/koril-israel-korea/. [14]

Nilsson, A. et al. (2021), *Green Hydrogen Applications in Mongolia: Technology potential and policy options*, New Climate Institute, https://newclimate.org/sites/default/files/2021/10/NewClimate_Green_Hydrogen_Applications_in_Mongolia.pdf. [28]

OECD (2024), *OECD STIP Compass: GERD as a percentage of GDP*, OECD, https://stip.oecd.org/stats/SB-StatTrends.html?i=G_XGDP&v=3&t=1998,2021&s=CHN,EU27_2021,JPN,KOR,OECD,USA,GBR. [33]

OECD (2023), *FDI Regulatory Restrictiveness Index*, OECD Publishing, Paris, https://stats.oecd.org/Index.aspx?DataSetCode=FDIINDEX. [68]

OECD (2023), *Policy Toolkit for Strengthening FDI and SME Linkages*, OECD Publishing, Paris, https://doi.org/10.1787/688bde9a-en. [79]

OECD (2023), *Risk-based Regulatory Design for the Safe Use of Hydrogen*, OECD Publishing, Paris, https://doi.org/10.1787/46d2da5e-en. [41]

OECD (2022), *Green Hydrogen Opportunities for Emerging and Developing Economies*, OECD, Paris, https://www.oecd-ilibrary.org/docserver/53ad9f22-en.pdf?expires=1696414509&id=id&accname=ocid84004878&checksum=01FB008F220DF39E4DDDC822F7FFEB07. [2]

OECD (2022), *OECD Reviews of Innovation Policy: Germany 2022: Building Agility for Successful Transitions*, OECD Reviews of Innovation Policy, OECD Publishing, Paris, https://doi.org/10.1787/50b32331-en. [20]

OECD (2021), *Recommendation of the Council for Agile Regulatory Governance to Harness Innovation*, OECD, Paris, https://www.oecd.org/mcm/Recommendation-for-Agile-Regulatory-Governance-to-Harness-Innovation.pdf. [42]

OECD/The World Bank/UN Environment (2018), *Financing Climate Futures: Rethinking Infrastructure*, OECD Publishing, Paris, https://doi.org/10.1787/9789264308114-en. [80]

Rodrik, D. (2014), *Green industrial policy*, Oxford Review of Economic Policy, No. Volume 30, Number 3, pp. 469-491, Oxford University Press, https://drodrik.scholar.harvard.edu/files/dani-rodrik/files/green_industrial_policy.pdf. [76]

Saygin, D. and M. Lee (2023), *Financing cost impacts on cost competitiveness of green hydrogen in emergin and developing economies*, OECD Environment Working Papers, No. No. 227, OECD, Paris, https://www.oecd-ilibrary.org/fr/environment/financing-cost-impacts-on-cost-competitiveness-of-green-hydrogen-in-emerging-and-developing-economies_15b16fc3-en. [29]

Tagliapietra, S. and R. Veugelers (2020), *A Green Industrial Policy For Europe*, Bruegel, https://www.bruegel.org/sites/default/files/wp_attachments/Bruegel_Blueprint_31_Complete_151220.pdf. [75]

Times of Oman (2021), *Oman's hydrogen alliance to drive national hydrogen economy*, Times of Oman, https://timesofoman.com/article/105391-omans-hydrogen-alliance-to-drive-national-hydrogen-economy. [58]

U.S. Department of State (2023), *2023 Investment Climate Statements: Mongolia*, Investment Climate Statements, U.S. Department of State, D.C., https://www.state.gov/reports/2023-investment-climate-statements/mongolia/#:~:text=Upon%2C%20Foreign%20Investment-,Policies%20Towards%20Foreign%20Direct%20Investment,investors%20face%20no%20investment%20minimums. [54]

UNDP (2023), *Mongolia SDG Investor Map: Report on Investment Opportunity Areas*, UNDP, Ulaanbaatar, https://investmongolia.gov.mn/download/sdginvestormap.pdf. [78]

World Bank (2023), *Advisory Report on the Development of a Green Hydrogen Certification Scheme in Chile*, World Bank, https://documents1.worldbank.org/curated/en/099062523123027598/pdf/P17631002007000f5080680467b6cdd230a.pdf. [52]

World Bank (2023), *Global Clean Hydrogen Financing Flagship Publication*, World Bank Group, DC. [18]

World Bank (2023), *Mongolia's Third Energy Sector Project: Project Information Document*, World Bank, DC, https://documents1.worldbank.org/curated/en/099062123053511517/pdf/P1781900f268cc010b6590592148fee6a1.pdf. [56]

World Bank (2023), *World Bank Development Indicators*, World Bank, DC, https://databank.worldbank.org/source/world-development-indicators. [34]

Notes

[1] IP5 patent families are patents filed in at least two offices worldwide, including one of the five largest IP offices: the European Patent Office (EPO), the Japan Patent Office (JPO), the Korean Intellectual Property Office (KIPO), the United States Patent and Trademark Office (USPTO), and the China National Intellectual Property Administration (CNIPA).

3 Aligning Investment Policy for Mongolia's Hydrogen ambitions: Renewable energy, low-carbon Technologies and Supporting Infrastructure

This chapter considers the investment policy framework for renewable hydrogen development in Mongolia. Experience from countries that have introduced national strategies for renewable hydrogen development underscores the importance of improving both the hydrogen-specific and general investment climates, as well as enhancing the supportive policy framework. As Mongolia considers its own national hydrogen strategy, the country has the opportunity to tackle some critical challenges to its investment attractiveness for clean energy and other low-carbon technologies. Key challenges include improving the attractiveness of the renewable energy sector relative to fossil fuel alternatives, laying the groundwork for the hard and soft infrastructure required to scale the industry, and stimulating demand for renewable hydrogen within Mongolian industry.

Introduction

There are two ways of looking at investment attraction in the renewable hydrogen sector in Mongolia. The first approach is to limit government intervention to establishing a conducive framework for local project development, through the identification of potential regulatory and policy barriers and hurdles and addressing them. The second – more directional – approach is to target the large-scale development of a national hydrogen sector through the mobilisation of massive public resources to accompany sectoral development. This second strategy combines elements of investment and 'green' industrial policy. Most economies establishing national strategies for the development of renewable hydrogen production, regardless of whether they envisage renewable hydrogen imports or domestic production, adopt the interventionist approach. Governments typically adopt this approach because achieving cost competitiveness of low-carbon hydrogen production, i.e., to reach prices on par with "grey" hydrogen, requires reaching economies of scale rapidly. While the break-even point is high in this capital-intensive sector, most low-carbon projects have not reached the stage of final investment decisions (FID) yet (IEA, 2023[1]).

The general investment policy framework is of structural importance since it enables or restricts private investment in general and foreign direct investment (FDI) in particular. In the case of the latter, there is a well-established literature on the role of FDI in enabling emerging technologies to move up global value chains and gain new knowledge and technological capacities. Specific, targeted renewable hydrogen investment policies include instruments such as incentives and financial schemes designed to directly support sectoral development, as well as dedicated investment promotion and facilitation efforts. Given that Mongolia will likely need to mobilise external finance if it is to achieve a hydrogen industry at scale, the chapter places particular importance on FDI. The chapter considers investment promotion policies across three categories, from the broadest to the most specific: Policies promoting private investment development in general; policies targeting green and low-carbon investments; and policies directed specifically at renewable hydrogen.

Whether Mongolia takes a more directional approach to promoting investment in hydrogen will depend on whether the country wishes to establish itself as a "first mover" hydrogen producing economy. The implications, both in terms of time and financial resources, will vary for investment promotion in hydrogen development depending on whether authorities in Mongolia decide to position the country as an early producer of a frontier, nascent technology, or whether it should wait for the global market, technologies, and infrastructure to mature before taking a more directional approach to the development of a hydrogen industry in Mongolia. Should Mongolia wish to establish itself as a first mover, then, as discussed in this chapter, there will likely be a need for substantial and well-coordinated investment policies that directly overcome and address investment risks relating to hydrogen-specific technologies, infrastructure and market.

At the same time, there is also scope for Mongolia to pursue a less directional approach to hydrogen investment, one which focuses on reducing barriers to and improving framework conditions for investment in low-carbon technologies more broadly. Such an approach would involve three primary areas for policy intervention. The first would be to ensure that the basic conditions necessary for private sector hydrogen development pilots in Mongolia to succeed are in place. The second would be to integrate low-carbon and environmental sustainability objectives into the investment policy framework. This approach would benefit the development of a hydrogen industry in Mongolia without overly committing the government to one technology option. Here the rationale is similar to the industrial and innovation policy discussion set out in Chapter 2: If the economic and social benefits of 'green' investments are not fully internalised by investors in the public and private sector, largely due to the absence of fully priced environmental externalities, then fossil-fuel-based technologies will remain more economically competitive than their green alternatives. Well-designed and targeted investment policy interventions can therefore help to address this imbalance. By combing the approaches outlined in this chapter, Mongolia could create an investment toolkit for its hydrogen industry that is in line with the country's economic situation and the government's priorities.

The third and critically important consideration for investment in Mongolia's hydrogen industry, whether the government takes a more directional approach to hydrogen or not, is the need for massive investment in renewable energy. A stable and low-cost supply of renewable electricity is necessary to competitively produce renewable hydrogen, and Mongolia's renewable energy potential is the country's key comparative advantage as a potential hydrogen producer. For hydrogen to be considered 'green' or 'renewable', the electricity used to power electrolysis must come from renewable sources. At present, electrolysers require 45-83 kilowatt hour (kWh) to produce 1 KG of H2, depending on the technology used (the more commonly used alkaline electrolysers have a range of 50-78 kWh/KG H2, whereas solid oxide electrolysers are more efficient with a range of 45-55 kWh/KG H2) (OECD, 2022[2]). Most renewable hydrogen studies assume that for renewable hydrogen to be competitive, projects will require a levelised cost of electricity (LCOE) of around 0.02 USD/kWh for wind and solar electricity generation. In reality, the global average RE generation costs are generally higher, with new utility-scale solar PV costing, on average, 0.048 USD/kWh and 0.033 USD/kWh for wind in 2021 (IRENA, 2022[3]). To achieve a utility-scale hydrogen industry in Mongolia, the country will need to significantly increase its renewable electricity generation capacity - just one of the pilot projects currently in Mongolia anticipates a need for an additional 600 megawatt (MW) of combined wind and solar power, more than double the 283 MW of renewable power currently installed in Mongolia as of 2024 – and ensure that the cost of renewable power generation is sufficiently low for competitiveness. Thus, the conditions for fostering investments in renewable energy in Mongolia will play a significant role in enabling a growing trajectory of renewable hydrogen production moving ahead, which is why a part of this chapter is devoted to exploring these conditions.

The chapter investigates the importance of investment policies for renewable hydrogen development in Mongolia and identifies measures and instruments that could be included in its national strategy. The chapter is structured as follows. Section 3.2 starts by looking at international experience in developing investment plans and policies for renewable hydrogen development in the framework of national hydrogen strategies. Section 3.3 then provides a view of Mongolia's renewable energy sector potential as the key source for domestic production competitiveness. Section 3.4 focusses on the assessment of the investment policy framework in Mongolia and its alignment with renewable hydrogen development. Section 3.5 concludes with a discussion on the importance of soft and hard infrastructure as enabling conditions for attracting the investments needed for the long-term industry development.

Building an investment strategy for hydrogen development: international experience

Policy discussions about renewable hydrogen largely focus on the need to engage in massive efforts to scale-up production and achieve cost-competitiveness of this technology versus its "brown" or "grey" alternatives and enable the reaching of net-zero targets by mid-century. Against this backdrop and considering their anticipated domestic needs for renewable hydrogen and potential economic and future low-carbon benefits from this nascent technology, many countries have developed renewable hydrogen national strategies with a strong emphasis on investments, sometimes within broader hydrogen strategies that also consider other production pathways (e.g., fossil-fuel based with CCUS). These strategies typically aim for specific target volumes of production, products and derivatives, sectoral markets, and target production costs. The economic literature on low-carbon hydrogen indicates that production costs should reach as low as USD 1 to 3/kg for renewable hydrogen to become competitive against "grey" hydrogen, from USD 3/kg for best-in-class projects to more than USD 10/kg in less favourable locations and conditions (OECD/The World Bank, 2024[4]). Countries also often provide estimated investment sizes associated with their visions. The examples provided below in Table 3.1 show a range of estimated cumulated investments ranging between USD 77 Billion and USD 330 Billion, which includes primarily private investment.

Table 3.1. Examples of "first-mover" renewable hydrogen exporting countries' investment strategies: targets and planned estimated investment

	Canada	Chile	Morocco	Namibia	Oman
Production targets (2030)	4 Mt	25 GW of electrolysis capacity	Export: 10.3 TWhrs Industry: 3.1 TWhrs	1-2 Mt	1-1.25 Mt
Production targets (2050)	20 Mt	N/A	export: 114.7 TWhrs industry: 20.7 TWhrs	10-12 Mt	7.5 - 8.5 Mt
Clean hydrogen products	Low-carbon H2	Green H2, green ammonia, e-methanol, e-gasoline, green steel	Green H2, green ammonia, green methanol	E-kerosene, green ammonia, green H2, hot-briquetted iron, methanol	Green H2, green ammonia
Sectors of application	Oil and natural gas industry, transport, heating, export	Power, mining, export	Fertilizers, transport, power, export.	Export, power	Refinery, export
Targeted production costs (2030)	USD 1.50-3.50/kg	USD 0.95-1.05/kg	N/A	USD 1.20-1.30 /kg	USD 1.60 /kg
Estimated cumulated investment	N/A	USD 330B*	USD 77B	USD 190B by 2040	USD 140B by 2050

Note: (*) Estimated cumulated investment opportunities for the private sector by 2050.
Source: Desk research.

Countries' national hydrogen strategies can be categorised into three types: importers' strategies, exporters' strategies, or self-sufficient strategies. Germany, Japan, Korea, and the EU, for example, have developed strategies for importing renewable hydrogen, anticipating their needs for reaching their low-carbon targets. China is a typical example of a self-sufficient renewable hydrogen producer, having a large domestic market able to absorb a scaled production of renewable hydrogen across a wide range of sectoral applications. Other countries have developed exporters' strategies, which, given Mongolia's small domestic market size, is a more appropriate model for an ambitious strategy, as discussed in Chapter 1. Canada, Chile, Morocco, Namibia and Oman's provide examples of exporters' strategies. These countries offer a range of various experiences of interest as countries with ambitions for becoming competitive renewable hydrogen exporters, illustrating what the competition looks like for an economy with similar ambitions.

The magnitude of investments needed to scale-up the sector is driving important policy discussions about mobilising financing for de-risking investments undertaken by the private sector and the role that international financial institutions should play in providing such financing in emerging economies. Recognising the limited ability of many developing economies to provide the public financial support needed to de-risk the development at scale of the sector, international partners are mobilised for directing financing at clean hydrogen development in economies endowed with natural assets and favourable conditions for domestic production at scale. The World Bank, in a report developed with the OECD, proposes a global initiative for supporting mid- to large size "lighthouse" projects (100 MW to 1GW) with a total volume of 10 gigawatts (GW) or more. The aim is to demonstrate the viability of renewable hydrogen projects across emerging economies. The report details further aspects of this initiative (Box 3.1) (OECD/The World Bank, 2024[4]). Other multilateral development such as the European Bank for Reconstruction and Development (EBRD)can provide financing in well-designed renewable hydrogen projects as part of their green investment portfolio. Understanding the requirements and criteria for accessing financing solutions offered by International Financial Institutions (IFIs) and Multilateral Development Banks (MDBs) directed at renewable hydrogen development should be a priority for emerging economies aiming to become major producing countries.

Box 3.1. Aspects of the World Bank's lighthouse project

- Develop a guideline or checklist of factors across the supply chain that need to be considered when developing a clean hydrogen project;
- promote tools for prioritizing and assessing projects, drawing initially from the pipeline of existing projects;
- identify and remove bottlenecks;
- enhance project quality to improve bankability;
- promote joint mechanisms (funds and donors such as MDBs) to allocate financial support to the projects;
- follow the principles of the World Bank Group's International Finance Corporation (IFC or similar environmental, social, and governance guidelines) to define a set of actions to guarantee sustainable, socioeconomically equitable, and beneficial project development.

Source: (OECD/The World Bank, 2024[4])

Exporters' national renewable investment strategies rely on domestic assets and competitive advantages for renewable hydrogen production and exports. National investment strategies of "first-movers" rely on identified national comparative advantages. For example, Chile relies on its low levelized cost of electricity produced from renewable sources to become one of the most competitive hydrogen exporters. Morocco and its partners highlight the country's potential for solar and wind energy development, as well as its proximity to EU markets and access via maritime routes. Namibia has some of the world's highest potential for solar and wind generation (NewClimate Institute, 2023[5]). Available infrastructure for hydrogen transportation, such as pipelines and ports, that can be retrofitted or upgraded, is also a key advantage. Australia's and Chile's port infrastructures provide these countries with good access to many markets via maritime routes, with regions as far as the EU considering them as potential renewable hydrogen suppliers. Finally, national investment strategies often plan to leverage the demand in domestic industries. For example, Chile plans to use renewable hydrogen in its large mining sector, while Morocco aims to use green ammonia in its fertilisers sector in substitution of imported grey ammonia. These domestic applications for renewable hydrogen produced locally are often meant to support the first stages of national hydrogen strategies. These applications require much less transportation infrastructure than exports and enable the creation of national pilot projects to demonstrate viability.

Emerging economies positioning themselves as first-movers in the renewable hydrogen space often have already large-scale investment projects under development with the involvement of international partners. Several emerging countries developing renewable hydrogen exporting strategies have strong backing from international private investors and foreign governments for developing large-scale projects and key infrastructure. Namibia, for example, has signed several Memorandums of Understanding (MoUs) with the European Commission, Belgium, Germany, and Japan. The government has signed on the commissioning of a giga-scale renewable project, Hyphen Namibia, representing USD 10 Billion of CAPEX and a full annual production capacity of 350 KT / year (NewClimate Institute, 2023[5]). The project is supported by two foreign private investors including a large German energy industrial. Other notable examples, closer to Mongolia, are the Svevind project in Kazakhstan (which released its national hydrogen strategy in May 2024), and ACWA Power's hydrogen project in Uzbekistan (which does not have a national hydrogen strategy at the time of the writing of this report).

Mongolia's renewable energy capacities and their importance for hydrogen development

The main comparative advantage of Mongolia as a producer country of renewable hydrogen is its renewable energy generation potential, the overwhelming majority of which has yet to be developed. The country is estimated to have 2.13 TW of wind power capacity with an electricity production capacity of 2,600 terawatt hours (TWh). This is equivalent to around 25% of total Chinese electricity consumption in 2022, and substantially above Mongolia's 7.27 TWh of electricity consumption in 2022 (IRENA, 2016[6]) (IEA, 2023[7]). Of the total potential, around 400GW are considered to be feasibly exploitable, with these located in the south and southeast of the country (World Bank, 2020[8]). The actual production capacity is, however, likely to be much greater. The most authoritative international study of Mongolia's wind resources remains the 2001 assessment conducted by the US National Renewable Energy Laboratory (NREL).[1] Whilst the NREL findings in terms of capacities and their location in Mongolia remain accurate, wind turbine generation capacity has significantly increased since the report's publication over twenty years ago. At that time, the average wind turbine was mounted at a height of 40 metres and had a capacity of 500 kW. In contrast, as of 2023, the typical height is around 100 metres and generation capacity of 3-4 MW (IRENA, 2023[9]).

The most productive wind areas are concentrated in the South Gobi region of Mongolia (Figure 3.1). Wind resources are most significant in the Omnogovi Aimag, Mongolia's largest region, with a generation capacity is around 300,000 MW (IRENA, 2016[6]). The neighbouring regions of Bajankhongor, Dundgovi and Dornogovi Aimag – where the country's largest wind farm, Saishand, is located, with 55 MW of installed capacity – also have high levels of wind generation capacity. This concentration indicates some of the country's most potentially productive wind power areas are concentrated in the south of Mongolia, relatively proximate to major industrial sites and Ulaanbaatar. The south and southeast of the country are also home to the regions with the highest wind capacity factors (CF), where CF ranges between 17.5-31.6%. Some thirteen aimags have more than 20,000 MW of wind potential, while nine have more than 50,000 MW of wind potential (IRENA, 2016[6]). Whilst there may be limited commercial rationale for large scale wind installations in all of Mongolia's aimags, the quality of wind resources throughout much of the country suggests that there is significant scope for using wind for off-grid power generation in remote areas that are otherwise difficult and costly to connect to the national grid.

Figure 3.1. Wind energy potential in Mongolia

Mongolia's wind resources are concentrated in the south and east of the country

Source: (World Bank/Global Wind Atlas, 2023[10])

As of 2023, Mongolia had 155 MW of installed wind power capacity. The current installed wind capacity accounts for10% of Mongolia's total installed power capacity and produced 564 gigawatt hour (GWh) in 2021 according to IRENA (although the IEA reports a lower level of 480GWh in 2021, falling to 420GWh in 2022). This output represents 70.4% of renewable electricity produced in 2021 and or 7% of total generation (IEA, 2023[7]) (IRENA, 2022[11]) (Ministry of Energy of Mongolia, 2022[12]). The current level of installed wind capacity is low, and while an additional 104 MW of wind power has been installed since the introduction of the government's 2015 *State Policy Paper on the Energy Sector*, there has been almost no new installed capacity since 2018 (New Climate Institute, 2020[13]). The government does have plans for the installation of an additional 252 MW of wind energy before 2030 (Government of Mongolia, 2023[14]).

Mongolia's solar energy capacities are also significant and are concentrated in the south and east of the country. Mongolia averages between 270-300 days of sun per year, with an estimated 2,250-3,300 hours of daylight. The country benefits from a high level of regular solar irradiance, with a daily solar energy capacity in the range of 3.4-5.4 kWh/m2 over an area of 23,461 square kilometres. Taken together, Mongolia has a hypothetical solar electricity generation capacity of 4,774 TWh per year, and an estimated exploitable solar capacity of 700GW. As with wind, the most productive solar regions are in the south and east of the country. The South Gobi region has particularly significant solar power generation potential, with the NREL calculating that Mongolia had an average solar electricity generation potential of 1,092 TWh over an area of 5,542 square kilometres (Figure 3.2).

Figure 3.2. Photovoltaic (PV) power potential in Mongolia

Mongolia's PV power is concentrated in the southern aimags of the country

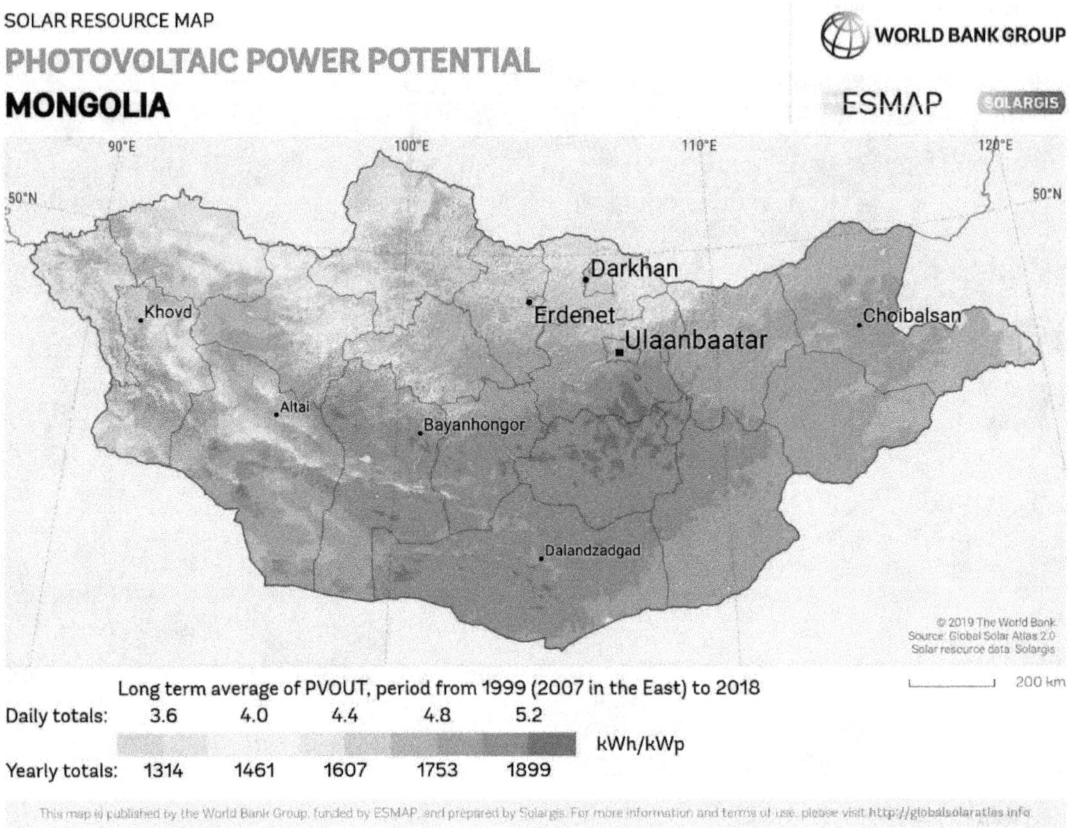

Source: (World Bank/Global Solar Atlas, 2023[15])

Despite Mongolia's significant solar resources, actual levels of installed solar power capacity remain low. Mongolia currently has 90 MW of installed PV power generation capacity, with an additional 35 MW capacity planned by the Ministry of Energy (Ministry of Energy of Mongolia, 2022[12]). There are also a number of projects in the west of the country to improve the efficiency and reduce transmission loss of existing installations. Despite the limited contribution to grid decarbonisation and renewable energy expansion, solar power has nevertheless played a transformative role in providing reliable electricity to rural communities.. Since 2002 and with the assistance of the World Bank, Mongolia implemented a massive rollout of over 100,000 small-scale solar power units with an annual generation capacity of 5,000 kW. This initiative has provided electricity to about half of Mongolia's population, or some 70% of the country's nomadic population (IRENA, 2016[6]).

Despite facing significant water scarcity issues, Mongolia has substantial hydropower potential, with these primarily concentrated in the north of the country. Annual precipitation is generally low, diminishing digressively from the mountainous north and northeast of the country close to the border with Russia towards the steppe and desert regions of the south and west. Precipitation is also highly seasonal, with there being almost no rain in winter months when much of the country's waterways and lakes are frozen. A detailed discussion of Mongolia's water resources and their suitability for sustainable, low-carbon hydrogen development is discussed in Chapter 2.

Hydropower accounts for the largest share of planned additional renewable electricity capacity. Of the 360 MW non-fossil fuel additions to the country's energy system planned by the Mongolian Ministry of

Energy, 315 MW is allocated to the Eglin Gol (EG) River Hydropower Plant (Ministry of Energy of Mongolia, 2022[12]). The EG plant, whose construction began in 2016 with USD 1 billion loan from China, will be Mongolia's first major hydropower installation, The project, however, has attracted controversy due to its potential impact on biodiversity and the environment, particularly on Lake Baikal in Russia, into which Mongolia's northern rivers drain (IRENA, 2016[6]). Ongoing discussions between Russia and Mongolia, intermediated by UNESCO, have been centred around the project's potential transboundary impact.. The UNESCO World Heritage Centre has requested additional information on the transboundary Strategic Environmental Assessment for the EG project, as well as for the Regional Environmental Assessments for the Shuren and Orkhon hydropower projects. As of summer 2024, the Mongolian authorities have not yet provided this information (UNESCO, 2023[16]).

Utilising Mongolia's potential hydropower resources is complicated by two geographical constraints. Unlike Mongolia's wind and solar resources, there is a geospatial mismatch between the location of the country's potential hydropower generation in the north and the demand for the electricity that could be produced, with Ulaanbaatar and the main centres of industrial demand in the centre and south of the country. The limited interconnectedness between Mongolia's five grid systems and a high level of transmission lost renders delivering power from the north to these locations expensive and inefficient under current infrastructure conditions. In addition, the development of hydroelectric power plants in the country's northern regions is complicated by potential objections from Russia, which has expressed concern over the impact of such projects on Lake Baikal, as outlined above (IRENA, 2016[6]).

Plans for overcoming the challenges of renewable electricity development in Mongolia will help provide a conducive environment for investments in renewable hydrogen. Although pilot projects in Mongolia currently envisage off-grid renewable electricity production for powering electrolysers, the prospect of selling renewable electricity production in excess can contribute to improved business models therefore encouraging investments even in the short term. In the longer-term, should Mongolia envisage the development of renewable hydrogen at scale, the volumes of electricity needed will require access to a decarbonised grid.. The section below considers the investment policy framework for investing in renewable energy production, which will directly impact future prospects for renewable hydrogen investment.

Assessing Mongolia's investment regulatory and policy framework for renewable hydrogen development

Investment attraction and incentives for renewable hydrogen development

As in other countries of Central Asia, Mongolia has a relatively open statutory framework for investment. The OECD FDI Regulatory Restrictiveness Index, which measures the statutory restrictiveness of different economic sectors for foreign investment, shows that Mongolia's economy has a similar level of FDI regulatory restrictiveness as the OECD average (Box 3.3) (Fig. 3.3). This is reflective of a relatively broad openness to foreign investment and the investment-friendly policies of the government, as noted in other international assessments (U.S. Department of State, 2023[17]). The challenge for Mongolia is that this *de jure* openness to foreign investment does not equate to a *de facto* openness. Non-statutory barriers that are adjacent to the legal framework for investment continue to weigh on the investment attractiveness, including that of clean energy and low-carbon opportunities, in the country. As in other Central Asia countries, the implementation of regulations in a way that ensures consistency, transparency and reliability remains a concern. Anecdotal evidence from the mining sector, a prominent FDI recipient and an engine of national growth, hints that foreign investors are facing potential legal challenges when developing significant projects in the country.

Figure 3.3. FDI Regulatory Restrictiveness Index: Mongolia, Central Asia, and the OECD

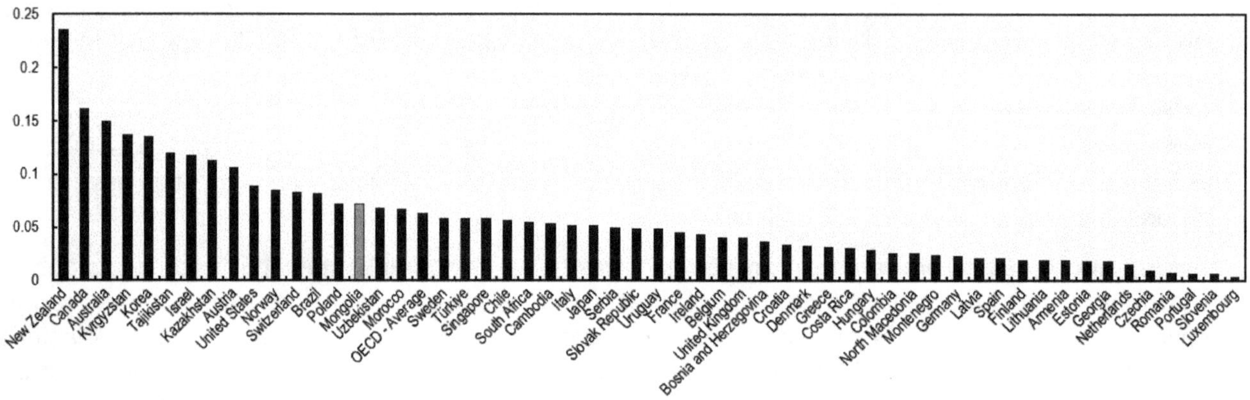

Note: 1 = most restrictive, 0 = least restrictive.
Source: (OECD, 2023[18])

While there are no explicit regulations or preferences for low-carbon technology projects that can support the decarbonisation of hard-to-abate industries, the government has begun to put together a legal framework that gives certain preferences and incentives to renewable energy projects. The 2013 Law on Investment plays a crucial role in this aspect, providing certain tax-free incentives for investment in innovative projects, as outlined in Article 12 of the law, though the scope of such projects is not defined; As of 2024, the Law on Investment was being revised by the Government of Mongolia and remained open to public comment (Government of Mongolia, 2023[19]). These incentives encompass a broad range of direct and indirect support measures on a number of key areas of licensing and permitting, such as long land tenure agreements (up to 60 years) with the possibility of further extensions (up to an additional 40 years), fiscal incentives for innovative export-oriented projects, support with visa and work permit requirements, and assistance with access to free economic zones and technology parks (Government of Mongolia, 2013[20]). As noted on the information portal of the newly-created Mongolian Investment Promotion Agency (IPA), the Law on Investment also establishes that investment projects that support the development of renewable energy and innovation capacities can also apply for stabilisation certificates, whereby certain taxes (CIT, customs duties, VAT, minerals royalties) can be held at fixed rates over a legally determined period of time (5-18 years) depending on the size of the investment (PwC, 2023[21]).

Mongolian legislation also allows for a number of preferential customs arrangements for the importation of goods for innovation and renewable energy infrastructure. As enshrined in Article 1 of the Exemption from Customs and Value Added Tax Law, imported equipment necessary for the development of innovative products will not be subjected to custom and value-added tax. At present, these exemptions do not extend to capital goods necessary for renewable hydrogen production. For instance, the import of Solar PV panels is exempt from a 5.5% customs tax. The government also exempts certain goods from customs and VAT, specifically targeting renewable energy research and production equipment, its accessories, spare parts, and equipment necessary for the production of innovative products. (Government of Mongolia, 2012[22]) (Government of Mongolia, 2008[23]). The Law on Energy Saving introduces fiscal incentives for constructing energy-efficient buildings and for manufacturing or importing energy-efficient machines and materials. Meanwhile, the Law on Corporate Income Tax, particularly under Article 22.5.3, offers tax relief for activities that efficiently use natural resources and reduce environmental pollution and waste. This includes the sale of environmentally friendly techniques and equipment, thus promoting greener industrial practices (Government of Mongolia, 2015[24]).

Box 3.2. The OECD FDI Regulatory Restrictiveness Index

The OECD FDI Regulatory Restrictiveness Index covers 22 sectors, including agriculture, mining, electricity, manufacturing and main services (transport, construction, distribution, communications, real estate, financial and professional services). Restrictions are evaluated on a 0 (open) to 1 (closed) scale. The overall restrictiveness index is a weighted average of individual sectoral scores.

For each sector, the scoring is based on the following elements:

- the level of foreign equity ownership permitted;
- the screening and approval procedures applied to inward foreign direct investment;
- restrictions on key foreign personnel; and
- other restrictions such as on land ownership, corporate organisation (e.g., branching).

The measures taken into account by the index are limited to statutory regulatory restrictions on FDI, typically listed in countries' lists of reservations under Foreign Trade Agreements (FTAs) or, for OECD countries, under the list of exceptions to national treatment. The FDI Index does not assess actual enforcement and implementation procedures. The discriminatory nature of measures, i.e., when they apply to foreign investors only, is the central criterion for scoring a measure. State ownership and state monopolies, to the extent they are not discriminatory towards foreigners, are not scored. Preferential treatment for special-economic zones and export-oriented investors or for investment from a treaty-covered country is also not factored into the FDI Index score.

Source: (Kalinova, Palerm and Thomsen, 2010[25])

In 2023, the government established a new Investment Promotion Agency (IPA). The agency, which operates under aegis of the Ministry of Economic Development, has a user-friendly website that lists sectors into which the government is actively seeking foreign investment. These include metal and mining, logistics, renewable energy, and agriculture. The site provides a clear overview of regulations relevant to each sector. It is difficult to attest to the effectiveness of the IPA due to the limited time it has been operational. However, considering the government's renewable hydrogen ambitions and the importance of attracting FDI. Furthermore, the IPA may be able to play an active role in soliciting potential hydrogen investors and providing information to those active or exploring opportunities in the country regarding the regulations, incentives and procurement opportunities available for renewable hydrogen – and other low-carbon – opportunities in Mongolia. There may be additional opportunities to strengthen the IPA's activities and the government's broader investment attraction ambitions through a closer alignment with similar activities undertaken by development partners, such as the UNDP's "Mongolia SDG Investor Map" (UNDP, 2023[26]).

Mongolia's recent efforts in developing a national long-term low-emission development strategy (LT-LEDS), if successful, can accelerate investment in clean energy and low-carbon technologies. The adoption of robust LT-LEDS with credible associated commitments has the power of sending out signals and incentives encouraging stakeholders to invest in low-carbon projects supporting the economic transformation necessary to achieve Paris Agreement targets. LT-LEDS and their implementation roadmaps typically contain sectoral contributions to the overall economies' decarbonisation, ideally with clear quantitative targets and identified supporting measures. If credible, i.e., backed with strong public commitments, such as financial and non-financial incentives, carbon pricing instruments and other public investment in areas critical to the transition, these strategies can play a catalytic role in the development of project pipelines supporting sectoral decarbonisation targets (OECD/The World Bank/UN Environment,

2018[27]). Mongolia has started to work with international partners to develop a LT-LEDS, and while this work is at a very early stage, its potential for attracting sustainable investment in the country is great, under the condition that the plan is credible, and that the government commits to its realisation.

Similarly, the development of a credible national renewable hydrogen strategy can contribute to attracting investment in the sector. By developing a cross-cutting vision for the sector growth and committing to pursuing policies in support of projects development, including in key areas such as contributing to securing demand at the nascent stage, a national strategy – the very topic of this report - can make a difference and encourage investors. If the process is well-established and if the strategy contains well-conducted economic analyses, it can also help investors develop business plans. Given that renewable hydrogen is a nascent technology, the development of credible plans, economic analyses and forecasts relies on a process that is consultative and well-documented. This is why, when developing national strategies, countries typically establish steering platforms bringing together representatives of various sectors who can inform a sound and credible strategy and oversee its implementation (Box 3.2).

Box 3.3. Co-ordinating low-carbon innovation and its deployment: practices in national renewable hydrogen strategies

The fast adoption and deployment at scale of an innovative low-carbon technology such as renewable hydrogen requires a high degree of commitment and co-ordination across stakeholders. Many countries have established institutional bodies with the mandate to oversee the development and implementation of national hydrogen strategies. These bodies bring together a wide array of stakeholders, from ministers and government officials to representatives of industry and academia. Some countries have also decided to set up ad hoc bodies to manage specific aspects of the strategy implementation.

- Morocco has established a National Hydrogen Commission for its national hydrogen strategy implementation. The "Commission Nationale de l'Hydrogène" was created in 2019. Chaired by the Minister of Energy, Mines and Environment, the Commission is composed of representatives from four different Ministries, industrial organisations such as the Confédération Générale des Entreprises du Maroc, research centers and state-owned enterprises.

- Chile set up an Interministerial Council of Hydrogen (Consejo Interministerial de Hidrogeno Verde) which brings together multiple ministries as well as regional governors. The mandate of the Council is to review and assess progress made by the bodies and stakeholders in charge of the strategy implementation, including the Steering Committee (Comité Estratégico), established in July 2023 The Committee has as its main objective the drafting of the Green Hydrogen Action Plan 2023-2030 to facilitate the development of this nascent industry (Government of Chile, 2023[48])

- Canada has established a Steering Committee to oversee the overall implementation of its hydrogen strategy, as well as working groups dedicated to specific topics. The mandate of the Steering Committee is to implement and track the progress made. It is co-chaired by the Minister of Natural Resources and by the CEO of the Canada Hydrogen and Fuel Cell Association. In addition, 16 working groups work on the strategy's implementation, in specific areas such as Codes and Standards, Export, Mining, and Training.

- Oman's approach is strongly oriented towards public-private co-operation. Established in 2021 by the Minister of Energy and Minerals (MEM), the Hy-Fly Alliance gathers 13 public and private organisations (IEA, 2023[30]). The purpose of the Alliance is to facilitate the production and use of renewable hydrogen through investments, technologies, policies, and expertise across the whole hydrogen supply chain (Times of Oman, 2021[49])

- Namibia's Green Hydrogen Council was established in 2021 by the President of the Namibian Republic. Its eight members include five Ministers, the Central Bank governor, the economic advisor to the President and the CEO of Namibia's investment promotion agency (GH2 Namibia). In 2023 the Implementation Authority Office was established by the Ministry of Mines and Energy with the mandate to identify the key hydrogen projects to be developed and oversee the auction of state-owned land to ensure the government's participation in project development. (Government of Namibia, 2022[16]).

Investments in Mongolia's electricity sector and implications for renewable hydrogen

Mongolia's power network is split into five areas: the Western, Altai-Uliastai, Central Energy System (CES), Eastern and Southern power systems. The interconnectedness between these different areas is limited due to the significant distances between them and the low transmission capacities of power lines, which in turn limits the ability of the country to integrate its different regions into one system. The CES, which serves Ulaanbaatar and nearby settlements, is by far the largest, accounting for 80% of the country's power supply in 2022, with the base load primarily provided by seven Combined Heat and Power (CHP) plants (World Bank, 2022[28]); other systems have significantly lower capacities, with the Eastern Energy System catering for only 100 MW of electricity demand. A transboundary interconnection between the Russian power grid and the CES, Western, and Eastern power systems of Mongolia allow for power imports, the importance of which has grown in recent years as generation capacity in the country has struggled to keep pace with rising demand. Power sector transmission projects to improve the high voltage linkages between the south and the east to the main source of demand in Ulaanbaatar and mining sites are currently being supported by development partners such as the World Bank and the EBRD, with the latter working on a high voltage transmission connection between the renewable-rich region of the South Gobi and Ulaanbaatar, anticipating the potential role renewables could play in meeting increasing electricity demand from that region while keeping a stable baseload from the CES CHPs.

Demand for electricity is growing rapidly, primarily driven by the industrial and residential sectors. Between 2010 and 2021, electricity demand from the industrial sector grew 129%, or from 7,538 terajoules (TJ) to 17,246TJ. This makes the industrial sector by far the largest single electricity consumer in terms of absolute value and in additional demand. If the industrial electricity demand increases at the same rate between 2020-2030, then this would represent 30,712TJ by the end of the decade. This figure is 6,172TJ more than total electricity demand in Mongolia in 2020 (Figure 3.4). A significant part of industrial electricity demand comes from the mining sector, with industrial enterprises adopting a range of strategies to adapt to a lack of confidence in the Central power system to meet their needs. For example, the Oyu Tolgoi copper mine in the Southern Gobi region is powered by a power plant in northern China, while in consultations with the OECD, firms such as RioTinto (the Oyu Tolgoi operator) and the large Mongolian industrial firm MAK outlined their strategies to build their own co-located renewable energy infrastructure in order to ensure security of electricity supply. Another motivation for firm-level investment in co-located renewable energy infrastructure is to have certainty over the carbon intensity of their electricity consumption, something which, at present, the national grid operators are not able to provide.

Figure 3.4. Electricity demand in Mongolia by sector (TJ)

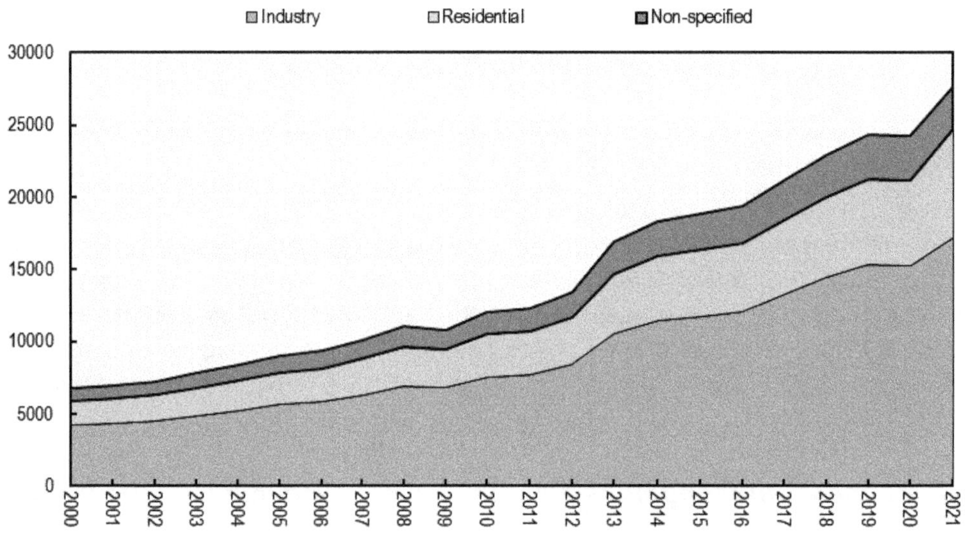

Source: OECD calculations based on (IEA, 2022[29])

In the face of rising electricity demand and limited domestic capacity, Mongolia has increasingly depended on expensive electricity imports. In 2020, Mongolia imported 1,705.6 GWh of electricity, or

19.3% of total domestic electricity consumption, of which all was imported from Russia – the only country with which Mongolia has a cross-border interconnection. Imported electricity creates significant costs for Mongolia, with the average import tariff in 2022 at 0.0928 USD/kWh, significantly higher than the export tariff of 0.0140 USD/kWh in the same year and almost double the domestic residential and industrial tariffs of 0.045/0.52 USD/kWh (IMF data). The inability of the grid to provide stable power has prompted some large industrial firms to begin investing in their own co-located renewable infrastructure.

The need for electricity imports stems from the inflexibility of Mongolia's power system, where electricity output generally follows heating demand, thereby causing electricity deficits at certain times of the year. The inflexibility of this system limits the ability of the grid to absorb even modest levels of electricity from variable renewable energy (VRE) sources (Nilsson et al., 2021[30]). Addressing the inflexibility challenges in the power system would allow Mongolia to integrate a substantially higher level of electricity generation from renewable sources, negating the need for additional coal-fired CHP expansions while eliminating its dependence on electricity imports to cover its power deficit. Sequencing policy decisions around hydrogen production, renewable generation development, and power system modernisation can all benefit the government's energy security objectives. Should the trend of industry co-locating their own energy infrastructure continue and be encouraged by the government – at present there is no clear policy support for these initiatives – it could decrease the strain on the central grid baseload and enable a larger share of renewable energy integration into the national power mix.

Electricity generation in Mongolia continues to be primarily coal powered. Given that renewable hydrogen requires electricity from renewable sources, the ability to draw on the national grid for renewable hydrogen generation in Mongolia is highly limited. In 2021, 84% of electricity generation came from coal (6269 GWh), with wind accounting for only 7.8% (580 GWh) and solar PV 2.2% (165 GWh). The predominance of coal in Mongolia's electricity generation mix means that, as power generation has grown to keep up with rising industrial demand there has been a precipitous rise in the carbon footprint of the energy sector. Emissions from burning coal for electricity (and heat) generation rose from 6.6 Mt of CO2 in 2000 to 14.5 Mt of CO2 in 2021 (Figure 3.5). Should electricity demand continue to increase at the same rate as in recent years and this increased demand be met by primarily coal-fired power plants, then it will be difficult Mongolia to meet its commitments to decarbonising the power sector. Additionally, this will have broader implications for the Scope 2 emissions (those related to industrial purchases of power and heat) for the expanding industrial sector (Box 3.1).

Figure 3.5. Electricity generation in Mongolia: source and emissions footprint

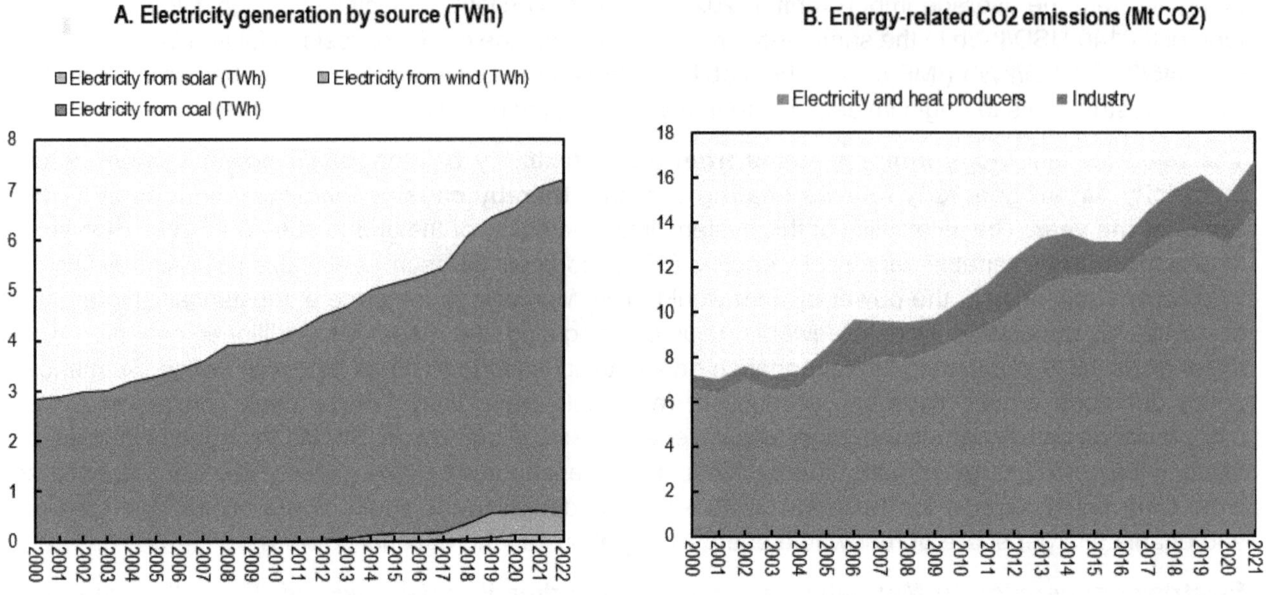

Note: Panel B displays combined heat and electricity CO2 emissions in Mongolia, reflecting the fact that all coal-fired power generation in Mongolia comes from Combined Heat and Power Plants (CHPs)
Source: OECD calculations based on (IEA, 2022[29])

Despite commitments to increase the share of renewable energy in the country's electricity generation, the majority of additional capacity is set to be coal-fired. The government has set clear targets for increasing the share of renewable energy in the country's electricity generation. The 2015 *State Policy on Energy* (SPE) set a goal of increasing renewable electricity capacity from 7.62% in 2014 to 20% by 2023, and then to 30% by 2030 as a share of total electricity generation capacity (IEA, 2023[7]). As part of the government's Mid-Term Development Plan of 2018 and in its Nationally Determined Contribution of 2019, Mongolia also set a goal of expanding hydropower capacity by 0.9 GW capacity from a level close to zero, as well as increasing renewable capacity by 0.6 GW from other sources (Government of Mongolia, 2019[31]) (Government of Mongolia, 2015[32]). In 2020, the government again increased its commitment under its Updated Nationally Determined Contribution targets to reduce GHG emissions from the energy sector by 49.4% by 2030 compared to its 2015 baseline. The government's state targets for renewable energy development nevertheless take place in a context where the bulk of anticipated electricity generation expansion will come from coal (Figure 3.5a) (Government of Mongolia, 2015[32]). The Ministry of Energy does anticipate the addition of 585 MW of renewable power by 2030, but these additions are dwarfed by the additional 2,090 MW of coal-fired power – including the 450 MW Tavantolgoi CHP – that is due to come online over the same period. If the capacity additions proceed as currently foreseen by the Ministry of Energy, the share of coal-fired CHP power in total power generation capacity will be greater in 2030 than it was in 2023 (89% vs. 83%) (Fig.3.3b).

Figure 3.6. Planned electricity generation expansion by source (2019-2030)

(a) MW, (b) as a percentage of total installed capacity

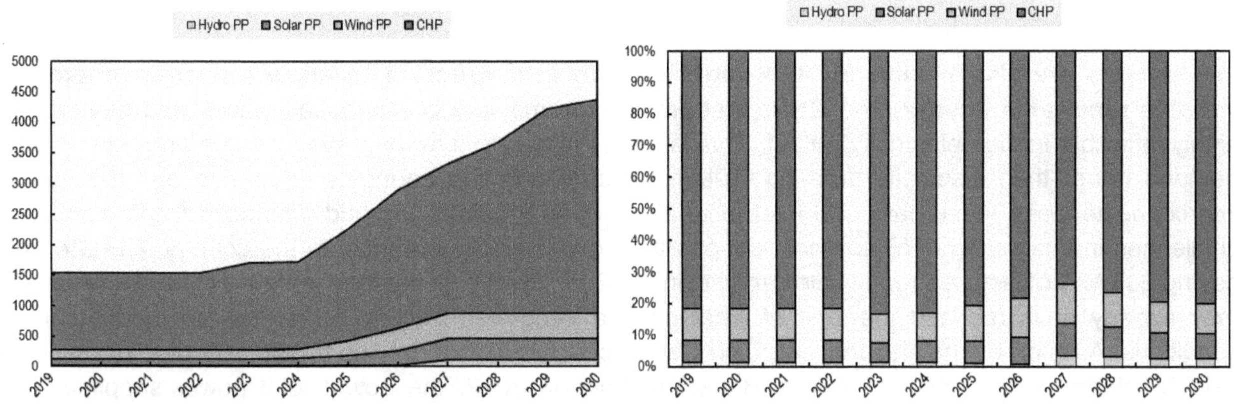

Source: OECD calculations based on information from the Ministry of Energy of Mongolia

Mongolia has the eighth largest pipeline of new coal power capacity in the world (Global Energy Monitor, 2021[33]). That such a significant proportion of additional power in Mongolia is due to come from coal rather than renewables puts the direction of development of the country's power sector at odds with the government's stated decarbonisation objectives, as well as with trends in global financing of energy infrastructure. For example, there is a significant gap between pre-construction projects (6,630 MW) and projects already under construction (200 MW) in Mongolia, which in part may reflect the increasing difficulties facing countries in attracting the investment and financing necessary for CHP development. In interviews with the OECD during the preparation of this report, stakeholders regularly cited the challenges the Mongolian government has faced in tendering CHP expansion projects. Due to the limited financial capacities of Mongolian State-Owned Enterprises (SOEs) to invest in energy infrastructure, Mongolia has been highly dependent on a limited number of bilateral (namely Japan and South Korea) and international financial partners to support new energy projects or the expansion of existing installations. However, most of these countries or institutions have either ceased financing new coal power or are in the process of ceasing such investments (Edianto, Trencher and Matsubae, 2022[34]). At the same time, developments such as the agreement of participants to the OECD Arrangement on Officially Supported Export Credits to end support for unabated coal-fired power plants is an indication that Mongolia's coal-driven power sector expansion will be increasingly difficult to finance in the years ahead (OECD, 2021[35]).

The lack of a clear plan to phase out unabated coal-fired power generation is a major barrier to developing Mongolia's renewable energy sector and incurs significant costs. The challenges in transitioning away from coal are far from unique to Mongolia. Supporting emerging economies – particularly those with large extractive sectors – in designing strategies and toolkits for such a transition has been the subject of work at the OECD (OECD, 2022[36]). Coal phase-out is often as much a political economy challenge as it is an economic or environmental one, with a multitude of complex policy issues interacting with one another: SOE governance issues in coal mining and power generation companies, the importance of taxes levied on production and trade of coal for government revenues, the role of low-cost electricity in maintaining a social contract between the state, its citizens and business, etc. Given the importance of coal for domestic power but also for export revenues, it is perhaps unsurprising that the IEA assessed Mongolia to be one of the six countries in the world where coal phase-out is likely to be the most challenging (IEA, 2022[37]). While there are undoubtedly costs involved in supporting the transition away from coal use, particularly in a country with abundant and cheap coal resources, it is important to note that there are equally costly implications – financial, social and environmental – in not doing so. Whether through the direct subsidy of coal use (e.g., the major coal-producing SOE Baganuur JSC providing state-owned power plants with coal below market prices, or charging below cost prices to electricity end-users) or indirect subsidies (e.g., soft budgetary constraints for SOEs in coal production and use), the business as usual situation in Mongolia directs relatively scarce public – and at times private resources – towards the continued use of coal. These resources could otherwise be used to accelerate the deployment of low-carbon alternatives (EITI, 2021[38]).

Investments are also necessary in Mongolia's grid infrastructure to allow for larger shares of variable renewable energy. The CHPs that provide the majority of Mongolia's power and heating are highly inflexible in their electricity output. This is primarily due to their operational focus on meeting heat demand rather than electricity demand. Taken together with the country's large size and the limited interconnectedness – internally and internationally – of its transmission grid, Mongolia faces significant challenges in integrating VRE sources, despite their considerable potential in the country. Despite only having 283 MW of installed renewable capacity in 2022, or 18.41% of the total installed capacity, Mongolia may already have reached the limit of what can be integrated into its power system under current conditions. A report by the National Dispatch Centre concluded that only up to 250 MW of VRE could be reliably integrated into the grid (World Bank, 2022[28]). In 2022, the actual total power supplied from renewable sources was equal to 10.24% of the total supply, or 0.64 TWh compared to 6.27TWh from fossil-fuel sources. Work supported by the EBRD is underway to improve grid transmission infrastructure between the renewable-rich South Gobi region and Ulaanbaatar. This connection would allow additional power demand in Ulaanbaatar to be provided by renewable energy generation in the south but would not necessarily displace coal-fired power in the capital.

Box 3.4. Understanding Scope 1, 2, and 3 emissions

Policymakers, regulators, firms, and international organisations regularly differentiate between the origin of emissions by referring to the 'scope'. Understanding this categorisation is key for measuring, inventorising, and addressing GHG emissions, as it is for navigating international and national regulations around GHG emissions. For example, in the EU Carbon Border Adjustment Mechanism (CBAM), the implementation rollout is explicitly linked to different scopes of emissions. The scopes of emissions are numbered, and broadly relate to upstream emissions, direct emissions at the activity being measured, and downstream indirect emissions:

Scope 1: This category covers direct GHG emissions that occur from sources that are owned or controlled by an entity (for example, emissions from the use of coal in a steel furnace, or from the vehicles used on an industrial plant). In 'hard-to-abate' sectors, it is often scope 1 emissions that are the most difficult and expensive to address.

Scope 2: This category covers GHG emissions from the generation of electricity purchased and consumed by the entity. Scope 2 emissions occur at the facility where the electricity is generated, not at the end-user.

Scope 3: These emissions are those that result as a consequence of the activities of an entity, but which occur from sources not owned or controlled by the entity. An example would be the emissions that occur through the production or extraction of materials used as inputs by the entity. Scope 3 emissions are particularly important for firms that are exploring how to improve the sustainability of their value chains.

Source: (Greenhouse Gas Protocol/World Resources Institute and World Business Council for Sustainable Development, 2004[39]) (European Commission, 2024[40])

Improving the attractiveness of Mongolia's clean energy and renewable hydrogen sector

In consultations conducted for this report, two key sets of issues emerged regarding the attractiveness of Mongolia's renewable energy sector. The first of these relates to market regulation, namely electricity tariffs. The second of these relates to investor confidence and public governance, which may have an impact on the willingness of foreign investors to enter into contracts with the Mongolian government or SOEs in the energy and power transmission sectors. The investment attractiveness of the country's renewable energy sector has implications for the country's low-carbon transition. Many factors influencing investment decisions to support this transition – including renewable hydrogen – are similar to those in the renewable energy sector. These factors encompass: entering into power purchasing agreements and other off-take contracts with the government or publicly owned entities, regulatory stability, and the broader regulatory framework for low-carbon investments. Such factors signal to investors that the government is serious about transitioning the economy to a less carbon-intensive footing.

As in other countries of Central Asia, Mongolia has a relatively open statutory framework for investment in the energy and electricity sector. As indicated in the OECD FDI Regulatory Restrictiveness Index, which measures the statutory restrictiveness of different economic sectors for foreign investment, Mongolia's electricity sector is generally less restrictive for FDI than the OECD average (Fig. 4.5). This is reflective of a relatively broad openness to foreign investment and the investment-friendly policies of the government, as noted in other international assessments (U.S. Department of State, 2023[17]). The challenge for Mongolia is that this *de jure* openness to foreign investment does not equate to a *de facto* openness. Non-statutory barriers, which operate alongside the legal framework for investment continue to weigh on the investment attractiveness of low-carbon opportunities, whether in energy or elsewhere, in the country.

Figure 3.7. OECD FDI Regulatory Restrictiveness Index: Primary sector, mining, electricity generation, electricity distribution

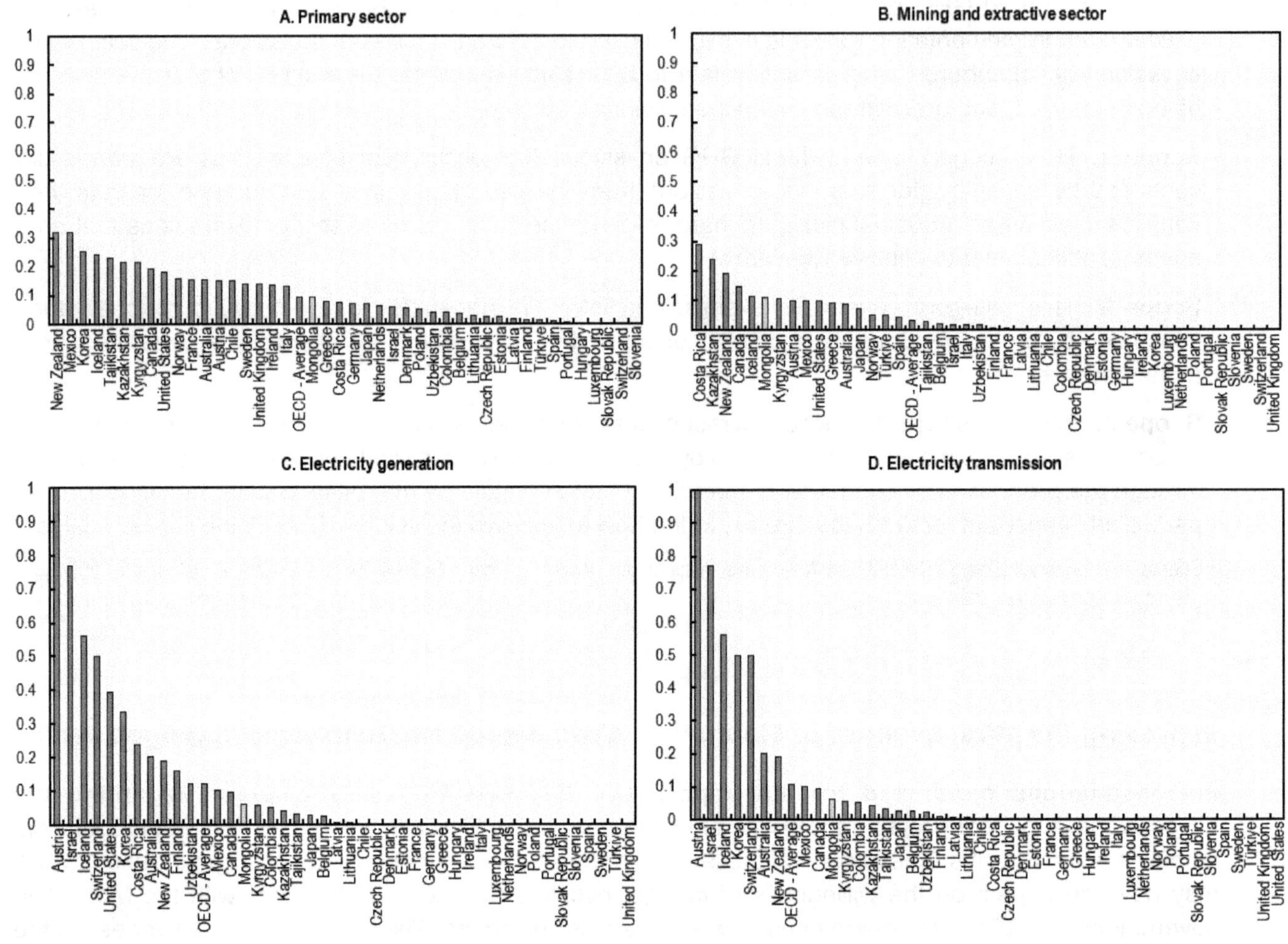

Note: Red = OECD average, orange = Mongolia
Source: (OECD, 2022[41])

Despite the statutory openness of power and energy sectors, Mongolia has struggled to attract FDI into renewable energy projects. In 2023, 74% of FDI stock in Mongolia was in the mining sector, with only 0.3% accounted for by activities related to the power sector (National Statistics Office of Mongolia, 2024[42]). While there are no official FDI statistics available specifically for renewable energy, Mongolia has succeeded in attracting 200 MW of private sector led renewable energy projects to the country (ADB, 2023[43]). The bulk of this investment took place following the introduction of feed-in tariffs in 2007, and much of it was co-financed by international donors, including the EBRD and World Bank. All of this investment has been in the Central Energy System, with the majority attributed to one project, the Sainshand Wind Farm in Dornogovi aimag. Since that project fully came online in 2019 there has been almost no additional external investment in the sector. Financing constraints for Mongolian SOEs in the power generation and transmission sectors mean that it is difficult for them to undertake the capacity expansion and network modernisation investments. These updates are necessary to meet rising demand

and policy commitments to energy sector decarbonisation detailed elsewhere in this chapter. The corollary of this is that, in the absence of reforms to electricity tariffs and a clear policy commitment to developing renewable energy and less carbon-intensive industrial production, attracting external finance – both from IFIs and the private sector – necessary for these investments may prove difficult.

Whereas financing for renewable energy projects is sensitive to local framework conditions, the same is not necessarily true for export-oriented extractive sector investment projects. Investment projects in the extractive sectors are often implemented by large, well-resourced multinational companies, many of which can negotiate directly with the government to overcome policy and regulatory barriers. Unlike Small and Medium Enterprises (SMEs), which are very sensitive to local framework conditions in the business environment, larger firms in extractive sectors can often self-finance, provide infrastructure, and ensure security. This ability enables them to operate in poor-quality institutional and legal environments. The implication is that success in the attraction of FDI into extractive sectors might hide underlying issues in the business climate that might limit success in attracting FDI into other sectors of the economy. Similarly, while statutory openness may be a necessary condition for FDI in non-extractive sectors, it may not be sufficient (OECD, 2021[44]). In the case of renewable energy, and indeed renewable hydrogen, issues that relate to investor confidence in the broader business climate and governance are particularly important. This is because such projects often require some kind of commercial agreement with the government or state-owned entity, at least in the early stages of project development, to make it to a final investment decision (World Bank, 2023[45]).

Regulated electricity prices are too low to attract the investment necessary to develop renewable energy projects to supply the national grid without additional policy intervention. Electricity prices in Mongolia are very low, with average residential tariffs and industrial tariffs of 0.045 and 0.52 USD/kWh. There is a balance to be struck between ensuring equitable access to electricity for the population and ensuring that tariffs are economically viable for investment in renewable energy. At present, the low cost of electricity is a function of abundant and cheap coal, a subsidised and primarily state-owned mining, generation, and distribution sector. There are ongoing discussions at the Energy Regulatory Commission and relevant line ministries, with input from the EBRD and other development partners, on a possible reform of the country's electricity tariff system, as well as on the introduction of competitive auctions for new renewable energy project developments.

The government has made limited use of financial de-risking instruments to improve the attractiveness of renewable energy projects for foreign investors. The main tool used by Mongolian energy regulators has been the feed-in tariff (FiT). Introduced in 2007, the Renewable Energy Law sets out the FiT ranges for different renewable technologies depending on whether they are on-grid or off-grid installations. These FiTs are paid in Mongolian Tugrik, contracts are signed with the NDC, and the PPA agreement lasts for ten years. The on-grid FiT bands are USD 0.08-0.095/kWh for wind and USD 0.15-0.18kWh for solar PV (IEA, 2023[7]). As part of ongoing discussions to reform the FiT and the regulated energy price the government is considering the introduction of competitive auctions for new renewable projects. In the absence of new revenue generation for the energy sector through pricing reform there is limited scope for the government to increase the incentives available to advance its renewable energy development ambitions.

The impact of these incentives on FDI in the renewable sector has been limited. Following initial success in attracting investment to develop the country's first large-scale wind farm, FDI into the renewable sector has been very low. One reason for this may be the structure and duration of Mongolia's FiTs, and importantly the regulatory rate to which these projects would revert following the expiry of the initial PPA. For example, authors of a 2023 study on Mongolia's renewable potential calculated that, once exposed to the regulated electricity price set by the Energy Regulatory Commission (ERC), utility-scale wind farms in Mongolia were economically unprofitable (Harrucksteiner et al., 2023[46]); the assessment considered a ten-year on grid feed-in tariff of 95 USD/ megawatt hour (MWh) for wind farms and 150.18 USD/MWh for solar PV, which then would be reduced to USD 37.4 USD/MWh under the ERC regulated rate, but does not consider the possibility for a separate PPA. It is important to note that the main factor affecting the economic viability of RE projects may not be the initial FiT, but the very low regulatory rate to which renewable projects would revert. Raising this regulatory rate for electricity is subject to an ongoing reform process in the country.

Another policy used to raise the attractiveness of investment in renewables is to lower the competitiveness of polluting equivalents through carbon pricing and other forms of regulation. Mongolia has yet to develop a carbon tax and is at a very early stage in developing a regulatory framework that disincentivises the use of coal. This is an issue for the expansion of renewable power generation in Mongolia for hydrogen production. For one, it severely limits the ability of the national grid to play any role in a domestic hydrogen economy, since the power it provides is overwhelmingly fossil fuel powered. Even if hydrogen projects were able to access renewable power at the margins of supply, this would not be consistent with the international regulatory trends that demand additional infrastructure for hydrogen production. Further, the cost of renewable electricity is in large part a function of the relative costs of alternatives; as coal-fired power becomes more expensive, there emerges a greater rationale to invest in and lower the cost of clean alternatives.

The government has highlighted renewable energy as a priority sector for investment attraction. In 2023 the government established a new Investment and Trade Agency (ITA) under the Ministry of Economic Development. Mirroring the investment objectives set out in the New Recovery Policy and Vision 2050, the ITA prominently features the renewable energy sector as one of the priority areas for FDI in the country. The site sets out clear information on regulation, licensing, and the applicable feed-in tariffs. While the ITA website does not yet discuss hydrogen, it may be advisable for the government to use this platform as well as its trade missions to promote the country as a potential location for renewable hydrogen production. The same logic is true for promoting clearer opportunities around power-intensive industrial production and the scope for co-located renewable electricity generation.

An unresolved state-investor dispute in Mongolia may limit the ability of the government to attract investment into higher-risk projects. Following the introduction of FiTs for renewable energy producers in Mongolia in 2007, Mongolia succeeded in attracting a number of renewable projects, with these supported with a combination of private and development finance. In consultations with investors in the Mongolian energy sector and potential investors in a low-carbon hydrogen industry, stakeholders frequently evoked an unresolved dispute between one of these early renewable projects and the Mongolian National Dispatch Centre (NDC) as a major barrier to developing the country's renewable sector and entering into higher risk industrial development projects such as low-carbon hydrogen. The dispute centres on establishing government liabilities to pay for curtailed electricity production from a wind farm in the South Gobi region. The Fit requires the NDC to pay a significantly higher rate for wind power generated than the equivalent from the country's CHPs, even when the grid is not able to integrate the renewable power. The dispute is indicative of a number of challenges facing the government if it is to attract FDI for other renewable projects and low-carbon hydrogen development: investor confidence in the government as an off-taker is relatively low and this is affecting investor appetite for further renewable projects in Mongolia, FiTs for low-carbon energy or fuels must be coherent with public financial capacities such that investors can be confident in the government's ability to uphold contractual obligations. Moreover, underinvestment in adjacent sectors – in this case power transmission – place a de facto limit on the scope and ambition of low-carbon investment projects in the country.

One of the key factors that affect FDI entering low-carbon projects in emerging markets is the risk-related higher costs of capital. It is important to note the main factor affecting capital costs in emerging economies are generally country-specific risks. Mitigating these risks often involves focusing on enabling conditions, such as sovereign guarantees or specialised insurance targeting policy and political uncertainties. Factors such as financial market maturity, policy stability, and the regulatory landscape also affect capital costs. In short, the cost of capital for hydrogen projects is affected by many of the same factors that affect the investment climate in the country more broadly, and as such improving the attractiveness of Mongolia for low-carbon industrial investments must be viewed through the same prism of issues that affect the investment climate in general, rather than as a separate category (Saygin and Lee, 2023[47]).

There are policy levers that governments can use to lower capital costs for low-carbon projects. For example, specialised tax treatment can improve early-stage project cashflows, helping to de-risk the investments of first movers. In a number of jurisdictions, governments are pursuing additional de-risking policies to support investors by lowering capital costs. This includes providing regulatory clarity for hydrogen developers, implementing FiTs (primarily for the renewable energy infrastructure built for hydrogen production), and establishing auctions (Table 3.2). Many of these policies are also relevant to the mobilisation of innovation resources for low-carbon development as outlined in Chapter 2. An important consideration for Mongolia is that the fiscal capacities available to policymakers to provide capital grants to early-stage projects are very limited. Therefore, policymakers may need to use a more creative blend of investment and fiscal incentives to achieve a de-risking effect. Mongolia is likely to require the mobilisation of finance from IFIs and MDBs to fulfil the role that public sector finance might play elsewhere. Consequently, it is essential ensure that hydrogen projects under consideration for public support in Mongolia are coherent with the financing criteria of these institutions.

Table 3.2. Risk factors and possible de-risking mechanisms

Risk Factors	Examples of de-risking mechanisms
Uncertain market demand	Purchase obligations Public procurement
Limited off-takers	Long-term hydrogen purchase agreement (HPA) Partial risk/credit guarantees Export credit guarantees
Uncertainty on hydrogen price	HPA Partial loan guarantees
Lack of hydrogen market	HPA Guarantees of origin to strengthen market credibility
Political risk	Political risk insurance Partial risk/credit guarantees
Limited infrastructure	Hydrogen hubs/clusters

Source: Adapted from (Saygin and Lee, 2023[47])

Box 3.5. Examples of national initiatives to mobilise investment for renewable hydrogen

Governments with hydrogen strategies have undertaken direct financing mechanisms to support the development of hydrogen development and deployment. Grants, subsidies, and funds are examples of devising measures implemented by government to de-risk investment and reduce market uncertainty. While public financial support for project development may be necessary, policymakers should pay further attention to market creation and complement their actions with demand-side policies (IEA, 2023[1]).

- The **SDG Namibia One Fund was launched at COP27 in 2022.** This blended finance platform aims at facilitating the deployment of green hydrogen market in Namibia. The fund was established in June 2023 through a partnership between the government-owned Environmental Investment Fund of Namibia and two Dutch organisations Climate Fund Managers and Invest international. The fund, whose objective is to mobilise a total of USD 1 billion, has already secured an initial funding of USD 40 million, under the form of a grant provided by Invest International. (Climate Fund Managers, 2022[48]) (Environmental Investment Fund, 2023[49]).
- The Chilean government through its development agency Corfo has awarded **USD 50 million of public subsidies to six renewable hydrogen projects.** (Green Hydrogen Organisation, n.d.[50]). Furthermore, in summer 2023 the government launched a USD 1,000 million Fund for the Development of Green Hydrogen. The aim of the fund is to catalyze private investment for production and demand projects (Government of Chile, 2023[51]).
- **Canada has implemented the Clean Fuels Fund (CFF)**. The federal government in Budget 2021 has allocated CAD 1.5 billion to de-risk capital investment for clean fuel production facilities, including hydrogen (Government of Canada, n.d.[52]). In 2022 60 projects were selected to receive federal funding under the CFF which will amount to CAD 800 million (Government of Canada, 2022[53]).
- **In its national hydrogen strategy, Morocco envisages the use of export credit guarantees as a de-risking instrument**. Morocco has highlighted in its green hydrogen strategy the importance that de-risking mechanisms play in rendering the investment environment more appealing to foreign investors. In particular, the strategy mentions direct support in the form of public private partnership and preferential tax treatment, as well as export credit guarantees as a potential option to attract investment (Government of Morocco, 2021[54]).

Hard and soft infrastructure for renewable hydrogen development in Mongolia

Hard infrastructure: developing the supporting infrastructure for transporting and storing renewable hydrogen

Export markets are likely to be key to supporting a domestic hydrogen industry at scale in Mongolia, and this will require additional infrastructure considerations. While it is important for projects to secure domestic off-take agreements, all the private-sector led renewable hydrogen projects in Mongolia envisage eventually producing hydrogen for export. It is evident that Mongolia's potential as a competitive hydrogen producer relies on its natural assets. However, the ability to be a competitive exporter of hydrogen will also depend on transportation costs.

For Mongolia, given its landlocked geography and distance to seaports, the destination of hydrogen exports will likely be China. Many of the projects consulted for this report anticipate rising demand for low-carbon hydrogen in China from both existing applications (such as fuel cell vehicles) and from novel applications in industry, notably as a substitute for coal – Mongolia's most important export commodity – in northern China's steel industry. Notably, China is currently undertaking significant hydrogen-related infrastructure work at all stages of the value chain. For production, there is the example of a large-scale UNIDO-backed renewable hydrogen demonstration project in the city of Baotou, which aims to be producing 390,000 tonnes of green ammonia annually by 2025 (UNIDO, 2023[55]). Within Chinese industry there are indications of where future demand may come from, particularly in the steel sector, The completion of China's first hydrogen-powered shaft furnace at the Zhanjiang Steelworks, one of the country's major iron and steel producers, serves as a significant example (Sinosteel, 2022[56]). For transmission, China has included the construction of a 400km gas pipeline as part of the next "Implementation Plan for the Construction of a National Oil and Gas Network" (People's Republic of China, 2023[57]).

There are generally three main ways to export hydrogen: ship, truck, and pipeline. Each mode of hydrogen transmission has its own advantages and disadvantages and can be more or less competitive depending on the distance being covered. Given the distance between the likely production sites in Mongolia and the nearest major market in Northern China (~400km), the most competitive option for Mongolia's exports would most probably be via pipeline, either as a compressed gas, liquified or transformed into another energy carrier such as ammonia. Countries with established gas transmission infrastructure are well-positioned here since the retrofitting of gas pipelines for hydrogen transmission can be achieved at low-cost (below 0.1 USD/kg H2 for distances up to 500km compared to USD 0.1-1/kg H2 for new pipelines). In comparison, transmission by ship and truck is much more expensive, at USD 1-2/kg H2 for distances up to 500km by truck and upwards of USD 2/kg H2 by ship, though this generally only used for distances above 5000km (Hydrogen Council/McKinsey & Company, 2021[58]). In Mongolia, unlike in other countries with ambitions to be early hydrogen producers such as Oman, there is no transboundary gas infrastructure that could be retrofitted. There are ongoing discussions to construct a pipeline from Russia to China that would cross Mongolia, but the rationale for this pipeline is to deliver natural gas to China (Downs, Losz and Mitrova, 2024[59]).

While pipeline construction costs vary depending on a range of local factors, the investment needs will nevertheless be substantial. At present, no transmission or storage infrastructure exists in Mongolia suitable for the export of locally produced hydrogen or its derivatives, and a clear understanding of what the future infrastructure needs, their associated costs, and how this affects the competitiveness of Mongolia as a hydrogen exporter will all need to be carefully considered if the country moves forward with a national hydrogen strategy. A survey of hydrogen infrastructure literature suggests that the cost of new pipelines ranges between USD 1.93 million to USD 3.95 million per kilometre.[2] Taking the location of a future utility-scale hydrogen production facility to be located at one of the pilot project sites in the South Gobi, and assuming that in order for the hydrogen produced there to be exported there would be a need for a pipeline from the production facility to Baotou, the first major hub in China, then approximately 310km of new pipeline would need to be constructed. Without considering the storage requirements or any additional infrastructure for transforming hydrogen into alternative carriers, this would imply an investment need of between USD 598.3 million and USD 1.224 billion.

The cheapest way to deliver hydrogen to the end-user is to cluster production and transmission infrastructure with demand. While it is true that in the short- and medium-term there is very limited domestic demand for renewable hydrogen or its derivatives in Mongolia, the very high investment costs related to hydrogen storage and its transmission infrastructure creates a clear economic rationale. This justification guides the development of the country's hydrogen industry in such a way that combines the most competitive production and delivery costs for both external and eventual domestic users. This is the logic that underpins the significant attention paid by policymakers in OECD economies, which emphasises the importance of co-locating hydrogen infrastructure – along with other supportive infrastructure for industrial decarbonisation – within existing or planned infrastructure clusters. While the development of industrial clusters will likely be driven by the market, there is nevertheless much that the government can do to encourage that development in a way which supports both its hydrogen export and domestic decarbonisation objectives. Interventions can range from simply providing the private sector and project developers with better information and data on hydrogen production and consumption opportunities (Box 3.6), to considering infrastructure sharing and co-location benefits in decisions around permitting and licensing. What is clear is that the more the government can steer the development of hydrogen production in Mongolia with a clustering of demand, the more Mongolia will be able to improve the economics of hydrogen use and the development of its supportive infrastructure.

Box 3.6. AusH2 - Australia's Hydrogen Opportunities Tool

The Australian government put in place a geo-mapping tool to support its national hydrogen strategy. The tool, developed by Geoscience Australia offers extensive data useful to hydrogen project localisation and feasibility assessment.

Australia's Hydrogen Opportunities Tool consolidates and provides free access to geoscience data and tools for mapping and understanding hydrogen potential in Australia. Bringing together 7,000 datasets, it allows to visualise many factors affecting project viability including natural resources and geographic factors (e.g., land type, surface and groundwater), infrastructure (existing and prospect) including renewable energy production sites and transport, and other information such as administrative boundaries, and land status. It also allows to visualise hydrogen projects under development, under construction or operating in Australia. It allows to build scenarios, providing data about potential for production with or without infrastructure constraints, potential for CCUS development, presence of carbon storage sites, etc.

The tool also enables users to run hydrogen project feasibility assessments through its Hydrogen Economic Fairways Tool (HEFT).

In addition, this tool acts as an information portal for hydrogen project developers. It contains links to a variety of information about the hydrogen economy including the national strategy, market studies, and issues paper.

Source: (Australian Government - Geoscience Australia, 2024[60])

Soft infrastructure: standards and certifications for renewable hydrogen projects

Ensuring conformity with standards and certification in emerging hydrogen value chains will be crucial for an export-oriented renewable hydrogen production strategy. If Mongolia seeks to position itself within international emerging low-carbon production value chains, either as an exporter of green hydrogen or its derivatives, then it will also need to ensure that domestic policymakers and private sector actors are aligned with the international quality infrastructure requirements underpinning regional and global low-carbon hydrogen markets. In a number of recent OECD and IEA publications on supporting low-carbon hydrogen development a recurring recommendation is for policymakers to engage in international consultations on the development of these standards and certifications, the landscape of which remains fragmented but which will likely consolidate in the coming years around a common set of rules and standards (Table 1.1) (Cammeraat et al., 2022; International Energy Agency, 2023b).

The source of energy for hydrogen production plays an important role in aligning products with international standards and certifications. While the standards and certifications that will govern the international trade in renewable hydrogen are still being developed, it is already clear that a reliable, certifiably green source of energy to power electrolysis will be central to any hydrogen project's compliance with the rules that will govern international hydrogen trade. Ensuring that the Mongolian authorities can give foreign investors – particularly those interested in using local value chains in a potential export-oriented production strategy – certainty that their projects will be compliant with international standards. This assurance will be crucial to the government's ambitions to develop a utility-scale hydrogen industry. While pilot-stage projects may be less concerned with international standards and certifications, there is a need for the government to ensure that it is abreast of external regulatory developments and that it uses this knowledge to guide its policy interventions to support pilot projects and domestic innovation. If the government is to decide to commit to the long-term scaling-up of renewable hydrogen production in the country, putting in place this soft infrastructure will become particularly crucial.

Importing markets will contribute to standardising low-carbon hydrogen production regulations. The EU will likely play a significant role in shaping the development of international standards for hydrogen due to its importance as an early market for the low-carbon technology and its derivatives. Major European economies such as Germany and the Netherlands have plans to import a substantial amount of their renewable hydrogen demand as they seek to advance their domestic decarbonisation agendas. In June 2023, the European Commission introduced two Delegated Acts to officially define renewable hydrogen and related fuels, designating them as renewable fuels of non-biological origin (RFNBOs). Within this framework, the EU distinguishes between renewable hydrogen produced via electrolysis from renewable energy sources and low-carbon hydrogen, which relies on fossil fuels with carbon capture and storage (Table 3.3).

Table 3.3. Emissions requirements from selected hydrogen labels and programmes

Jurisdiction	Label	Year	Requirements	Purpose
European Union	Renewable fuel of non-biological origin	2022 (not yet adopted)	70% lower life cycle GHG emissions than an equivalent fossil fuel (3.4 kg CO2-eq/kg H2) delivered to a customer, or 80% after 2026 if used for electricity or heat. Only electrolysis hydrogen from renewables is eligible and must prove additionality.	Eligible fuels (including hydrogen- based fuels) can contribute to national renewables targets.
European Union	EU taxonomy aligned hydrogen	2022	73.4% lower life cycle GHG emissions than an equivalent fossil fuel (3 kg CO2-eq/kg H2 life cycle emissions at the point of production, or 70% for a hydrogen-based fuel).	Guide investments into climate change mitigation
United States	Clean hydrogen	2022	Less than 4 kg CO2-eq/kg H2 life cycle GHG emissions up to the point of production.	Eligibility for funding instruments, with lower emissions eligible for more support.
Victoria, Australia/Germany	Zero-carbon hydrogen	2020	Hydrogen comes solely from renewable sources.	To receive guarantees of origin from the Smart Energy Council and Hydrogen Australia.

Note: Adapted from IEA World Energy Outlook 2022
Source: (IEA, 2022c)

The first EU Delegated Act demonstrates the importance of additionality requirements. The additionality requirement means that renewable power used for hydrogen electrolysis will only be classed as 'renewable' if the power does not detract from the decarbonisation (electrification) of other economic activities (European Commission, 2024[61]). In short, it means that for hydrogen imported into the EU to be considered 'green', then the renewable power used to produce it should either come from newly installed renewable energy infrastructure to meet power demand from electrolysis or surplus renewable power from the grid. This is a point of critical importance for emerging economies that are still in an early stage of power sector decarbonisation. The limited integration of renewable power into national power grids makes it difficult to certify that the power used for electrolysis is green. Additionally, even if it is green, it might not qualify as such in EU legislation due to failing the additionality requirement.

Learning from international experience and standards is crucial for new market entrants, allowing them to avoid duplicating efforts and align with global best practices. Adopting internationally recognised certification schemes, particularly for countries aspiring to export green hydrogen or its derivatives, is one approach; these messages have been increasingly highlighted by the OECD and IEA (IEA, 2023[1]). In countries such as Mongolia, where projects are expected to rely heavily on external technology developments, policymakers may explore an "outside-in" approach to regulatory development. This involves external regulatory developments and trends guiding the development of the domestic regulatory framework for renewable hydrogen and other synthetic fuels (see, for example, the ongoing development around Namibia's Synthetic Fuels Act and its inclusion in the country's hydrogen action plan to 2025) (European Commission, 2023[62]; Government of Namibia, 2022[63]) (OECD, 2023[64]) (World Bank, 2023[65]).

The production, storage, transport, and use of hydrogen beyond its traditional uses in Mongolia will likely require a significant revision and update of industrial safety standards. Industrial regulations and inspections for hydrogen in Mongolia are overseen by the National Energy Centre (NEC), which sits under the Ministry of Energy. While the NEC has extensive experience with the use of hydrogen in its limited industrial applications in Mongolia (primarily as a coolant for CHPs), the novelty of its use in new industrial applications, along with innovations in renewable hydrogen production and related advancements in storage and transmission, implies that the standards and regulations governing hydrogen's use will likely require updates. In this regard Mongolia faces a challenge similar to many other countries. Significant work exists at the OECD on updating licensing, permitting, inspections and enforcement for hydrogen production and usage. Given ; these new applications of hydrogen are not foreseen in existing legislation, they are often by default unaccounted for or subject to requirements that are not linked to specific risks of its new applications. An extensive overview of the OECD regulatory landscape for hydrogen can be found in the 2023 OECD publication *Risk-based Regulatory Design for the Safe Use of Hydrogen* (OECD, 2023[64]).

References

ADB (2023), *Mongolia: Upscaling Renewable Energy Sector Project*, ADB, https://www.adb.org/projects/50088-002/main#project-pds. [43]

Australian Government - Geoscience Australia (2024), *AusH2 - Australia's Hydrogen Opportunities Tool*, https://portal.ga.gov.au/persona/hydrogen. [60]

Climate Fund Managers (2022), *Launch Of The New 'SDG Namibia One Fund' At Cop27 Targets The Development Of A Green Hydrogen Economy In Namibia*, https://climatefundmanagers.com/2022/11/09/launch-of-the-new-sdg-namibia-one-fund-at-cop27-targets-the-development-of-a-green-hydrogen-economy-in-namibia/. [48]

CMS (2021), *HYDROGEN LAW, REGULATIONS & STRATEGY IN CHILE*, https://cms.law/en/int/expert-guides/cms-expert-guide-to-hydrogen/chile. [76]

Cordonnier, J. and D. Saygin (2023), *Financing solutions to foster industrial decarbonisation in emerging and developing economies*, OECD Environmnt Working Papers, OECD, Paris, https://www.oecd-ilibrary.org/docserver/24a155ab-en.pdf?expires=1705514321&id=id&accname=ocid84004878&checksum=8B304BE24018556EA4E6F4666DCA9A93. [78]

Downs, E., A. Losz and T. Mitrova (2024), *The Future of the Power of Siberia 2 Pipeline*, Center on Global Energy Policy at Columbia University, https://www.energypolicy.columbia.edu/publications/the-future-of-the-power-of-siberia-2-pipeline/. [59]

Edianto, A., G. Trencher and K. Matsubae (2022), *Why do some countries recieve more international financing for coal-fired power plants than renewables? Influencing factors in 23 countries*, Energy for Sustainable Development, No. Vol. 66, Elsevier, https://www.sciencedirect.com/science/article/pii/S0973082621001484#bb0515. [34]

EITI (2021), *Subsidies at What Cost? Shedding Light on State Support for Fossil Fuel Consumption*, EITI, Oslo, https://eiti.org/sites/default/files/attachments/en_eiti_fossil_fuel_subsidies_policy_brief.pdf. [38]

Environmental Investment Fund (2023), *THE GOVERNMENT OF NAMIBIA AND DUTCH PARTNERS INK MEMORANDUM OF UNDERSTANDING*, https://www.eif.org.na/post/sdg-namibia-one-press-release-20-june-2023. [49]

European Commission (2024), *Carbon Border Adjustment Mechanism*, European Commission, https://taxation-customs.ec.europa.eu/carbon-border-adjustment-mechanism_en. [40]

European Commission (2024), *Renewable Hydrogen*, European Commission, https://energy.ec.europa.eu/topics/energy-systems-integration/hydrogen/renewable-hydrogen_en#:~:text=The%20idea%20of%20additionality%20is,unsupported%20renewable%20electricity%20generation%20capacity. [61]

European Commission (2023), *Global Gateway: EU and Namibia agree on next steps of strategic partnership on sustainable raw materials and green hydrogen*, European Commission, https://ec.europa.eu/commission/presscorner/detail/es/ip_23_5263. [62]

Global Energy Monitor (2021), *Boom and Bust 2021*, Global Energy Monitor, https://globalenergymonitor.org/wp-content/uploads/2021/04/BoomAndBust_2021_final.pdf. [33]

Goverment of Mongolia (1994), *Land Law*, Government of Mongolia. [68]

Government of Canada (2022), *Minister Wilkinson Announces up to $800 Million in Project Funding to Advance Canada's Clean Fuels Sector*, https://www.canada.ca/en/natural-resources-canada/news/2022/11/minister-wilkinson-announces-up-to-800-million-in-project-funding-to-advance-canadas-clean-fuels-sector.html. [53]

Government of Canada (2020), *Canada Hydrogen Strategy*, Government of Canada, https://natural-resources.canada.ca/climate-change-adapting-impacts-and-reducing-emissions/canadas-green-future/the-hydrogen-strategy/23080. [75]

Government of Canada (n.d.), *Clean Fuels Fund*, https://natural-resources.canada.ca/climate-change/canadas-green-future/clean-fuels-fund/23734. [52]

Government of Chile (2023), *Gobierno presenta Fondo por US$ 1.000 millones para el desarrollo del Hidrógeno Verde en Chile*, Government of Chile, https://energia.gob.cl/noticias/nacional/gobierno-presenta-fondo-por-us-1000-millones-para-el-desarrollo-del-hidrogeno-verde-en-chile. [51]

Government of Mongolia (2023), *Draft Revision on the Law on Investment*, Ministry of Economy and Development of Mongolia, Ulaanbaatar, https://investmongolia.gov.mn/the-updated-draft-of-the-investment-law-is-available-for-comments-and-recommendations/. [19]

Government of Mongolia (2023), *Planned Energy Projects*, Ministry of Energy, Ulaanbaatar. [14]

Government of Mongolia (2019), *Mongolia's Nationally Determiend Contribution to the United Nations Framework Convention onf Climate Change*, Government of Mongolia, http://www.en.ccrcc.mn/stat/file/u/NDC_1.pdf. [31]

Government of Mongolia (2015), *Energy Law*, Government of Mongolia. [71]

Government of Mongolia (2015), *Intended Nationally Determined Contribution of Mongolia*, Government of Mongolia, http://www.ccrcc.mn/stat/file/u/INDC_Full-Document_ENG_2016.pdf. [32]

Government of Mongolia (2015), *The Law on Energy Saving*, Government of Mongolia. [24]

Government of Mongolia (2014), *Law of Mongolia on Petroleum*, Goverment of Mongolia. [70]

Government of Mongolia (2013), *Investment Law*, Government of Mongolia. [20]

Government of Mongolia (2012), *Exemption from Customs and Value Added Tax*, Government of Mongolia. [22]

Government of Mongolia (2008), *On Customs Tariffs and Customs Taxes*, Government of Mongolia. [23]

Government of Mongolia (2006), *Minerals Law*, Govermnet of Mongolia. [69]

Government of Mongolia (2006), *Toxic and Hazardous Chemicals Law*, Government of Mongolia. [73]

Government of Morocco (2021), *FEUILLE DE ROUTE HYDROGÈNE VERT*, https://www.mem.gov.ma/Lists/Lst_rapports/Attachments/36/Feuille%20de%20route%20de%20hydrog%C3%A8ne%20vert.pdf. [54]

Government of Namibia (2022), *Green Hydrogen and Derivatives Strategy*, Ministry of Mines and Energy of Namibia, https://www.ensafrica.com/uploads/newsarticles/0_namibia-gh2-strategy-rev2.pdf. [63]

Government of Namibia (2022), *Namibia Hydrogen Strategy*, Governmen of Namibia, https://www.ensafrica.com/uploads/newsarticles/0_namibia-gh2-strategy-rev2.pdf. [74]

Green Hydrogen Organisation (n.d.), *Country Portal - Chile*, https://gh2.org/countries/chile#:~:text=Six%20projects%20by%202025,subsidies%20totalling%20US%2450%20million. [50]

Green Hydrogen Organization (n.d.), *Chile's green hydrogen vision*, https://gh2.org/countries/chile. [77]

Greenhouse Gas Protocol/World Resources Institute and World Business Council for Sustainable Development (2004), *The Greenhouse Gas Protocol: A Corporate Accounting and Reporting Standard*, Greenhouse Gas Protocol/World Resources Institute and World Business Council for Sustainable Development, https://ghgprotocol.org/sites/default/files/standards/ghg-protocol-revised.pdf. [39]

Gutschow, J. et al. (2022), *The PRIMAP-hist national historical emissions time series*, Earth Syst. Sci., No. Data, 8, pp. 571-603, https://doi.org/10.5194/essd-8-571-2016. [81]

Harrucksteiner, A. et al. (2023), *A geospatial assessment of the techno-economic wind and solar potential of Mongolia*, Sustainable Energy Technologies and Assessments, No. Vol. 55, https://doi.org/10.1016/j.seta.2022.102889. [46]

Hydrogen Council/McKinsey & Company (2021), *Hydrogen Insights: A perspective on hydrogen investment, market development and cost competitiveness*, Hydrogen Council/McKinsey & Company, https://hydrogencouncil.com/wp-content/uploads/2021/02/Hydrogen-Insights-2021.pdf. [58]

IEA (2023), *GLobal Hydrogen Review 2023*, IEA, Paris, https://www.iea.org/reports/global-hydrogen-review-2023. [1]

IEA (2023), *Mongolia*, Greenhouse Gas Emissions from Energy, IEA, Paris, https://www.iea.org/countries/mongolia. [7]

IEA (2023), *Renewable Hydrogen from Oman*, IEA, https://www.iea.org/reports/renewable-hydrogen-from-oman. [67]

IEA (2022), *Coal in Net Zero Transitions: Strategies for rapid, secure and people-centred change*, IEA, Paris, https://iea.blob.core.windows.net/assets/4192696b-6518-4cfc-bb34-acc9312bf4b2/CoalinNetZeroTransitions.pdf. [37]

IEA (2022), *IEA World Energy Balances 2022*, IEA, Paris, https://www.iea.org/data-and-statistics/data-product/world-energy-statistics-and-balances. [29]

IEA (2022), *Renewable Hydrogen frm Oman: A producer economy in transition*, IEA, Paris, https://iea.blob.core.windows.net/assets/338820b9-702a-48bd-b732-b0a43cda641b/RenewableHydrogenfromOman.pdf. [66]

IRENA (2023), *Wind Energy*, IRENA, https://www.irena.org/Energy-Transition/Technology/Wind-energy#:~:text=In%201985%2C%20typical%20turbines%20had,the%20length%20of%20its%20blades. [9]

IRENA (2022), *Renewable Energy Statistics 2023*, IRENA, https://mc-cd8320d4-36a1-40ac-83cc-3389-cdn-endpoint.azureedge.net/-/media/Files/IRENA/Agency/Publication/2023/Jul/IRENA_Renewable_energy_statistics_2023.pdf?rev=7b2f44c294b84cad9a27fc24949d2134. [11]

IRENA (2022), *Renewable Power Costs in 2021*, IRENA, Abu Dhabi, https://www.irena.org/-/media/Files/IRENA/Agency/Publication/2022/Jul/IRENA_Power_Generation_Costs_2021.pdf?rev=34c22a4b244d434da0accde7de7c73d8. [3]

IRENA (2016), *Renewable Readiness Assessment: Mongolia*, IRENA, https://www.irena.org/-/media/Files/IRENA/Agency/Publication/2016/IRENA_RRA_Mongolia_2016.pdf?rev=b7b8dc32d10745cc9b70b55d31569366. [6]

Kalinova, B., A. Palerm and S. Thomsen (2010), "OECD's FDI Restrictiveness Index: 2010 Update", *OECD Working Papers on International Investment*, No. 2010/3, OECD Publishing, Paris, https://doi.org/10.1787/5km91p02zj7g-en. [25]

Ministry of Energy of Mongolia (2022), *Heat Supply and Renewable Energy Policy Planning*, Ministry of Energy of Mongolia, Ulaanbaatar, https://www.irena.org/-/media/Files/IRENA/Agency/Events/2022/May/Heat-supply-and-renewable-energy-policy-planning.pdf?la=en&hash=FC805CF764BBD7F6A001D27B9FDE9FF4170A0B48. [12]

National Statistics Office of Mongolia (2024), *Foreign Direct Investment Stock by Economic Sector*, National Statistics Office of Mongolia, https://www.1212.mn/en/statistic/statcate/573075/table-view/DT_NSO_1500_005V3. [42]

New Climate Institute (2020), *The Mongolian electricity sector in the context of international climate mitigation efforts*, New Climate Institute, https://newclimate.org/sites/default/files/2022-03/decarbonization_pathways_mongolia_0.pdf. [13]

NewClimate Institute (2023), *The landscape of green hydrogen in Namibia*. [5]

Nilsson, A. et al. (2021), *Green Hydrogen Applications in Mongolia: Technology potential and policy options*, New Climate Institute, https://newclimate.org/sites/default/files/2021/10/NewClimate_Green_Hydrogen_Applications_in_Mongolia.pdf. [30]

OECD (2023), *FDI Regulatory Restrictiveness Index*, OECD Publishing, Paris, https://stats.oecd.org/Index.aspx?DataSetCode=FDIINDEX. [18]

OECD (2023), *Policy Toolkit for Strengthening FDI and SME Linkages*, OECD Publishing, Paris, https://doi.org/10.1787/688bde9a-en. [80]

OECD (2023), *Risk-based Regulatory Design for the Safe Use of Hydrogen*, OECD Publishing, Paris, https://doi.org/10.1787/46d2da5e-en. [64]

OECD (2022), *Equitable Framework and Finance for Extractive-based Countries in Transition (EFFECT)*, OECD, Paris, https://www.oecd-ilibrary.org/docserver/7871c0ad-en.pdf?expires=1709306320&id=id&accname=ocid84004878&checksum=78AA36C78DCA45C6AF7173404CF154F6. [36]

OECD (2022), *Green Hydrogen Opportunities for Emerging and Developing Economies*, OECD, Paris, https://www.oecd-ilibrary.org/docserver/53ad9f22-en.pdf?expires=1696414509&id=id&accname=ocid84004878&checksum=01FB008F220DF39E4DDDC822F7FFEB07. [2]

OECD (2022), *OECD FDI Regulatory Restrictiveness Index*, OECD, Paris, https://stats.oecd.org/Index.aspx?datasetcode=FDIINDEX. [41]

OECD (2021), *Agreement reached at OECD to end export credit support for unabated coal-fired power plants*, OECD, Paris, https://www.oecd.org/newsroom/agreement-reached-at-oecd-to-end-export-credit-support-for-unabated-coal-fired-power-plants.htm. [35]

OECD (2021), *Improving the Legal Environment for Business and Investment in Central Asia*, OECD Publishing, Paris, https://www.oecd.org/eurasia/Improving-LEB-CA-ENG%2020%20April.pdf. [44]

OECD (2021), *Recommendation of the Council for Agile Regulatory Governance to Harness Innovation*, OECD, Paris, https://www.oecd.org/mcm/Recommendation-for-Agile-Regulatory-Governance-to-Harness-Innovation.pdf. [72]

OECD (2015), *Policy Framework for Investment: 2015 Edition*, https://doi.org/doi.org/10.1787/9789264208667-en. [79]

OECD/The World Bank (2024), *Scaling Hydrogen Financing for Development*, OECD Publishing, Paris, https://doi.org/10.1787/0287b22e-en. [4]

OECD/The World Bank/UN Environment (2018), *Financing Climate Futures: Rethinking Infrastructure*, OECD Publishing, Paris, https://doi.org/10.1787/9789264308114-en. [27]

People's Republic of China (2023), *"West-to-East Hydrogen Tramission" Pipeline Included in National Planning*, GOV.CN, https://www.gov.cn/yaowen/2023-04/10/content_5750685.htm. [57]

PwC (2023), *Mongolia Corporate Tax Credits and Incentives*, PwC, https://taxsummaries.pwc.com/mongolia/corporate/tax-credits-and-incentives. [21]

Saygin, D. and M. Lee (2023), *Financing cost impacts on cost competitiveness of green hydrogen in emergin and developing economies*, OECD Environment Working Papers, No. No. 227, OECD, Paris, https://www.oecd-ilibrary.org/fr/environment/financing-cost-impacts-on-cost-competitiveness-of-green-hydrogen-in-emerging-and-developing-economies_15b16fc3-en. [47]

Sinosteel (2022), *China's First Million-ton Hydrogen-based Shaft Furnacce Kicks off at Baosteel Zhanjiang Steelworks*, Sinosteel, http://mecc.sinosteel.com/EN/news_content.aspx?id=3778. [56]

U.S. Department of State (2023), *2023 Investment Climate Statements: Mongolia*, Investment Climate Statements, U.S. Department of State, D.C., https://www.state.gov/reports/2023-investment-climate-statements/mongolia/#:~:text=Upon%2C%20Foreign%20Investment-,Policies%20Towards%20Foreign%20Direct%20Investment,investors%20face%20no%20investment%20minimums. [17]

UNDP (2023), *Mongolia SDG Investor Map: Report on Investment Opportunity Areas*, UNDP, Ulaanbaatar, https://investmongolia.gov.mn/download/sdginvestormap.pdf. [26]

UNESCO (2023), *45 COM 6B.24 Lake Baikal (Russian Federation) (N 754)*, UNESCO, Paris, https://whc.unesco.org/en/soc/4340. [16]

UNIDO (2023), *DEMONSTRATION PROJECT ON PRODUCTION OF GREEN HYDROGEN AND AMMONIA UNDERWAY IN BAOTOU, CHINA*, UNIDO, https://hydrogen.unido.org/news/demonstration-project-production-green-hydrogen-and-ammonia-underway-baotou-china. [55]

World Bank (2023), *Advisory Report on the Development of a Green Hydrogen Certification Scheme in Chile*, World Bank, https://documents1.worldbank.org/curated/en/099062523123027598/pdf/P17631002007000f5080680467b6cdd230a.pdf. [65]

World Bank (2023), *Global Clean Hydrogen Financing Flagship Publication*, World Bank Group, DC. [45]

World Bank (2022), *Mogolia Third Energy Sector Project (P178190)*, World Bank Group, DC, https://documents1.worldbank.org/curated/en/099062123053511517/pdf/P1781900f268cc010b6590592148fee6a1.pdf. [28]

World Bank (2020), *Mongolia InfraSAP: Infrastructure for Connectivity and Economic Diversification*, World Bank, DC, https://openknowledge.worldbank.org/server/api/core/bitstreams/08c3160c-51f3-520d-ac21-223780282464/content. [8]

World Bank/Global Solar Atlas (2023), *Global Solar Atlas: Mongolia*, World Bank, DC, https://globalsolaratlas.info/map?r=MNG&c=47.923705,102.260742,6. [15]

World Bank/Global Wind Atlas (2023), *Global Wind Atlas: Mongolia*, World Bank, DC, https://globalwindatlas.info/en/area/Mongolia. [10]

Notes

[1] The US National Renewable Energy Laboratory 2001 *Wind Energy Resource Atlas of Mongolia* provides an extensive overview of Mongolia's wind resources, their location, and their technical characteristics. It remains the reference point of work on the wind energy sector by organisations including IRENA and Mongolia's Ministry of Energy.

[2] The distance (in kilometres) was estimated using the coordinates of Sainshand Wind Farm and Barun Industrial Zone Damao Baotou. The cost per kilometre was approximated using comparable projects and findings published by the Hydrogen Council and McKinsey & Company (Hydrogen Council/McKinsey & Company, 2021[58]).

4 Access to water for renewable hydrogen production

A reliable supply of clean, high-purity water is a necessary input for the process of electrolysis that produces renewable hydrogen. Access to water is therefore a critical determinant of a renewable hydrogen project's economic viability and environmental sustainability. While Mongolia does have abundant clean and replenishable water, this is mostly located in the north of the country. This creates a geospatial mismatch with the availability of freshwater for electrolysis and the location of Mongolia's high-quality renewable energy resources. These resources are primarily located in the water-scarce South Gobi region of the country, where the growing mining sector has already put local aquifers under pressure. Any national hydrogen strategy will need to contend with how to sustainably manage Mongolia's water resources if the country is to become a competitive, utility-scale producer of renewable hydrogen. This chapter examines some of the key water-related considerations for policymakers in Mongolia in the context of the country's hydrogen ambitions.

Introduction

Hydrogen production via water electrolysis would need to reach 520 metric ton (MT) in 2050 according to the IEA's Net Zero Target. Governments, industrials representatives, international partners and other stakeholders are undertaking efforts to help accelerate the global development of renewable and low-carbon hydrogen, as illustrated in the COP28 declaration of intent on mutual recognition of certification schemes for renewable and low-carbon hydrogen and hydrogen derivatives. In the sole segment of electrolysis – the cleanest hydrogen production pathway if renewable electricity is used – installed capacity and number of announced projects are growing rapidly (IEA, 2023[1]). The global installed water electrolyser capacity for renewable hydrogen production reached almost 700 megawatts (MW) at the end of 2022 - a 20% year-on-year increase - and could reach more than 2,000 MW by the end of 2023. Based on announced projects, 175 gigawatts (GW) could be reached by the end of the decade (IEA, 2023[1]). The anticipated massive scale-up of this technology is prompting experts and policymakers to raise the question whether this can have negative consequences on water resources availability (Beswick, Oliveira and Yan, 2021[2]).

This concern resonates particularly strongly in Mongolia, as well as in other countries of Central Asia, which are landlocked and affected by water stress. Many countries envisaging to become leaders in renewable and renewable hydrogen markets, for example in Latin America or the Middle East, plan to desalinate seawater for renewable hydrogen production, an option that Mongolia and most of its Central Asia neighbours do not have. Water is generally a sensitive issue in the region, and Mongolia faces water stresses at local levels (Asian Development Bank, 2020[3]), sparking concerns among the population and within the government. The needs to develop the water infrastructure are considerable, and a report by the NewClimate Institute finds that the growth of a renewable hydrogen sector would be conditional to such development (Nilsson et al., 2021[4]) As a result, access to water is a critical question that policymakers and potential investors (and other stakeholders) are consistently raising as the development of renewable hydrogen production in Mongolia is envisaged.

This chapter aims to address concerns about water access for renewable hydrogen production in Mongolia, in line with the goal of providing insights to inform a future national hydrogen strategy. Firstly, it seeks to determine an order of magnitude of the water consumption of a potential future renewable hydrogen sector in Mongolia. This will be done by drawing on the existing literature and project data collected on the ground, comparing it with other water uses and identifying associated risks. Secondly, it investigates available solutions for tackling water access problems for hydrogen production – excluding seawater desalination since this is not a viable option for Mongolia – and considers their potential application in Mongolia. Lastly, it provides a brief assessment of Mongolia's water policy framework as it relates to renewable hydrogen development, identifying potential issues and remedies.

Water needs of a future renewable hydrogen economy in Mongolia: implications and risks

Water needs from a developing renewable hydrogen sector raises concerns, particularly in Mongolia and the Central Asia region. Access to water is critical to the production of renewable hydrogen, which is produced via water electrolysis. The electrolysis, moreover, requires water of a high purity, meaning that in most cases water that is withdrawn for the production of hydrogen needs to be pre-treated with advanced technologies to achieve the required feed water quality standard. As policymakers across economies are pushing for a rapid scale-up of renewable hydrogen given its criticality for achieving net-zero by mid-century, a number of stakeholders worldwide have raised their concern that a large and rapid development of a renewable hydrogen sector could put additional pressure on freshwater resources at global, national, and local scales. This concern resonates in dry, landlocked Mongolia, where water is seen by stakeholders interviewed for this report as the number one issue to be addressed when raising the question of developing a renewable hydrogen sector.

Assessing water-related needs of renewable hydrogen in Mongolia

Existing information about water needs for renewable hydrogen production is subject to a number of caveats. Recent studies have provided estimates of water needs of future renewable hydrogen sectors in different countries. This section draws on this literature to provide estimates for a Mongolian production of renewable hydrogen.

There are a few caveats to note when considering water withdrawals and consumptions for renewable hydrogen production. First of all, the scope of predicted water needs can differ between studies. While some argue that only water used for electrolysis should be considered (Beswick, Oliveira and Yan, 2021[2]), others take a more holistic view, taking into account the whole industrial process of hydrogen production (including water used for cooling purposes), and even water used for the manufacturing and operation of renewable energy equipment powering electrolysers, differentiating between wind and solar energy production needs (Tonelli et al., 2023[5]). Some renewable hydrogen projects in Mongolia, as elsewhere, are considering producing renewable hydrogen in order to then convert the molecule to an alternative energy carrier (such as ammonia) or to liquify it. These additional steps have their own water requirements, which are not captured in this chapter, but nevertheless remain salient to the discussion on developing a utility-scale hydrogen industry in the country.

Another important caveat is that the renewable hydrogen sector is nascent, relying on a new technology untested at large scale. Current water withdrawal and consumption estimates therefore rely on scarce information based on laboratory conditions (where tap water is used) rather than real-life conditions. While laboratory conditions are likely to be less favourable than real world conditions, the technologies being used are also likely to mature and become more efficient in the coming years (IRENA and Bluerisk, 2023[6]).

According to the most recent scientific and technical literature, producing 1 kg of renewable hydrogen requires an estimated 20 to 30 litres (L) of water. This includes approximately 10 L/kg for the electrolysis, and an additional 10 to 20 L/KG for purification treatment and cooling purposes. Again, specific estimates vary according to key assumptions and the scale of production (Table 4.1).

Table 4.1. Water needs estimates for renewable hydrogen production in recent literature

Water needs in kilograms for one ton of renewable hydrogen produced

	Total water withdrawal	**Total water consumption**	Withdrawal for water purification	Withdrawal for electrolysis	Process cooling	Comments
ACS Energy Letters	**9**	**9**	- (1)	9	-	Freshwater is used.
Tonnelli et al.	**24**	**-**	15	9	*Water demand increase estimated at +19%*	The study also considers water used for RE production.
IRENA – alkaline*	**32.24**	**22.28**		14.19(2)	18.05(2)	Tap water (pre-treated) is used. Evaporative cooling is assumed.
IRENA – PEM**	**25.70**	**17.52**		11.31(2)	14.39(2)	
Eurokhan case study(3)	**-**	**-**	-	12	-	-

Notes: IRENA provides ranges and averages, only averages are shown here; (1) In the case of seawater desalination the study considers that water withdrawal would be twice the water volume for consumption; (2) calculated based on ratios provided in the report. (3) See Annex A. Alkaline* and proton exchange membrane (PEM)** electrolysers are already commercially available technologies used for electrolysis. Alkaline electrolysers are a more mature technology and have been used for a long time in industry, though they have – like PEM electrolysers – been used only relatively recently for the dedicated production of hydrogen. Both technologies are classified as being at a mature stage of technological readiness (TLR9) by the IEA.
Source: (Beswick, Oliveira and Yan, 2021[2]) (IRENA and Bluerisk, 2023[6]) (Tonelli et al., 2023[5]), OECD.

The water used as a feedstock for the electrolysis process needs to be of high purity due to the requirements of the electrolyser's technology, thus requiring specific pre-treatment according to its origins and composition. Water impurities can affect electrolyser performance and lifetime as well as hydrogen quality (Becker et al., 2023[7]). Recovery rates – the amount of water that is recycled and reused of the total water that has been drawn – after water treatment depend on several parameters including type of water used as an input, electrolyser characteristics (including technology, alkaline or Proton Exchange Member (PEM)), and type of pre-treatment. The core of the water pre-treatment is reverse osmosis for demineralisation, which has a recovery rate typically comprised between 40% and 60%, although it can reach 90% when fully optimised (U.S. Department of Energy, 2013[8]). This means that the balance is rejected as a brine and requires disposal. On the other hand, raw water can be used for cooling purposes. Water used for cooling doesn't have the same pre-treatment requirements as feed water used for electrolysis, therefore it can be of a higher mineral content. This also means that different water networks are required for the hydrogen production process.

In the case of Mongolia, the planned source of water for renewable hydrogen projects is underground water available in the South Gobi Desert, which is currently the preferred location of project developers (with the exception of one blue hydrogen project, which intends to draw from surface water, see Annex A). While assessments have been made of the capacities of aquifers in the South Gobi, information collected for this report shows that there seems to be a general consensus that these assessments are insufficient and are in need of being updated and improved. In the absence of such updated assessments, it is difficult to assess and predict the impact of hydrogen production on these aquifers, and by extension, the implications of such production on the availability of water for other uses in the region, with a high level of accuracy. The groundwater in the South Gobi is saline, and this has implications for the amount of water that would need to be drawn, the cost of this water (additional cost of desalination), and the environmental impact of its use (for example, through the production of brine[1]).

Available information contained in the literature and provided by local project developers can give guidance on the order of magnitude of water consumption necessary for hydrogen production in Mongolia.

Figure 4.1. below provides a simplified view of water uses in the renewable hydrogen production process and gives estimates calculated for Mongolia based on information from local case studies and data provided in the available literature, notably in the IRENA study. Underlying assumptions include the use of groundwater, a requirement of 12 L of water / kg for the electrolysis reaction after treatment and an overall water recovery rate of 30%, as well as the use of alkaline electrolysers. Water used for renewable electricity production (manufacturing and operation of equipment) is not considered due to a lack of relevant data and negligible effect (IRENA and Bluerisk, 2023[6]).

Figure 4.1. shows that producing 1 kg of renewable hydrogen production in Mongolia requires to withdraw 41.45 L of raw water. This includes 18.24 L of water withdrawn for the pre-treatment and electrolysis (enabling to use 12 L of ultra-pure feed water for the reaction). Additionally, 23.21 L of raw water used in the cooling phase. Of the water used in pre-treatment, 6.24 litres are rejected as discharge water from reverse osmosis, or brine, which needs disposal. Most of the water used in the cooling process is consumed through evaporation, while 6 L are discharged as warm water, which can be recycled.

Figure 4.1. Simplified water chart for renewable hydrogen production in Mongolia

Water inputs and rejects estimates are provided for the case of Mongolia, in litres.

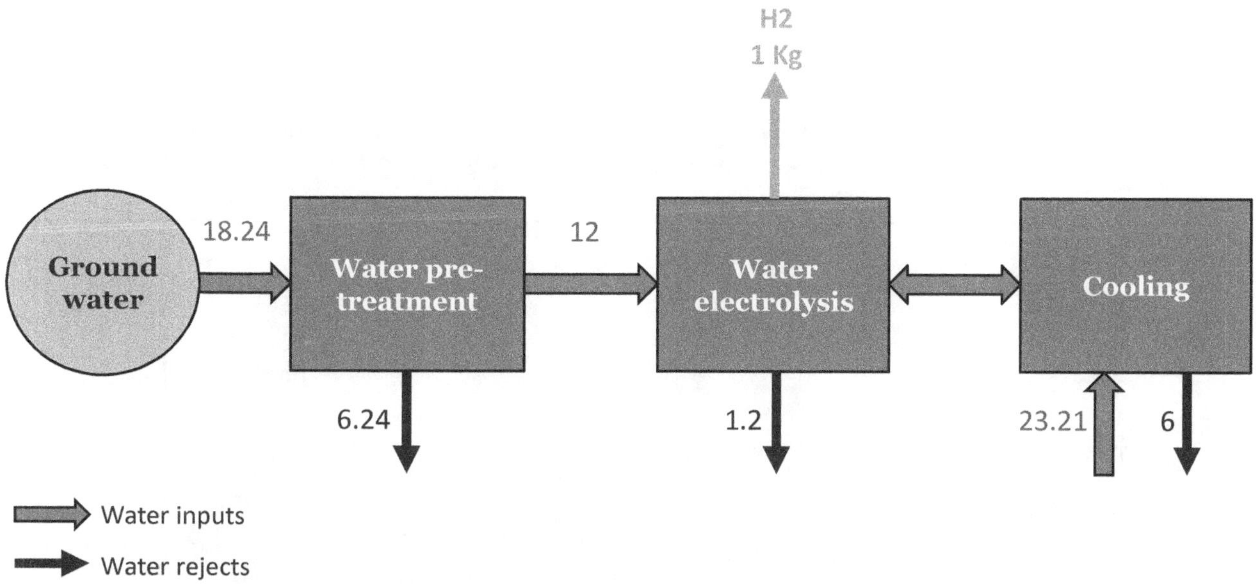

Source: Author based on IRENA, OECD case studies and Tonelli et al.

Project developers interviewed for this report (including some with projects outside of Mongolia) generally do not express concern that water requirements could impact the costs of renewable hydrogen production, specifically expressed as the levelized cost of hydrogen (LCOH), even when their projects are located in water stressed areas. Although water pre-treatment requires additional amounts of energy, capital and operational expenditure, it is considered that these represent a relatively negligible additional cost in the overall renewable hydrogen production cost structure (0.01 to 0.02 USD per KG H2 according to various sources), with water overall representing less than 1% of the total LCOH in cost competitive locations (Saygin and Lee, 2023[9]). Even in the case of seawater desalination where the water recovery rate is the lowest and energy needs are the highest, costs remain below 2% of the overall cost of hydrogen production[2]. Interviews with local project developers confirm that this is consistent with their own estimates taking into account the context of Mongolia. Hence, from the point of view of cost competitiveness, water treatment for renewable hydrogen production (including seawater desalination) is not considered pivotal.

Access to water, on the other hand, poses a critical issue. If the hydrogen industry were to bear the cost of transporting water to the production site, rather than relying on public investments in water infrastructure, it could result in a competitiveness issue due to the lack of adequate water infrastructure in Mongolia.

Based on the estimates provided in Figure 4.1, it is possible to simulate water withdrawals and consumptions for different capacities of a potential future renewable hydrogen sector in Mongolia, expressed in annual volume of production (Figure 4.2). In light of the limitations mentioned above, the focus is on the general scale rather than precise figures. The projected sector sizes are based on interviews with local project developers (local project - phases I and III) and on target objectives of reference countries seeking to position themselves as leading renewable hydrogen exporters. The size of the sector will be a key determinant of cost competitiveness due to its high capital intensity, including for the development of hydrogen storage and transport infrastructure. Mongolia aims to become an exporter of renewable hydrogen to China, which is already working to become a cost-competitive location for renewable hydrogen production, particularly as an electrolyser manufacturing country, one of the main cost determinants of production costs, alongside renewable electricity. Given this context, and Mongolia's consideration of its own national renewable hydrogen strategy, the estimates below offer a view of the water required if Mongolia were to match the ambitions of other leading hydrogen exporting countries.

Figure 4.2. Projected water demands from projects in pilot and scale-up phase

In thousand cubic meters / year

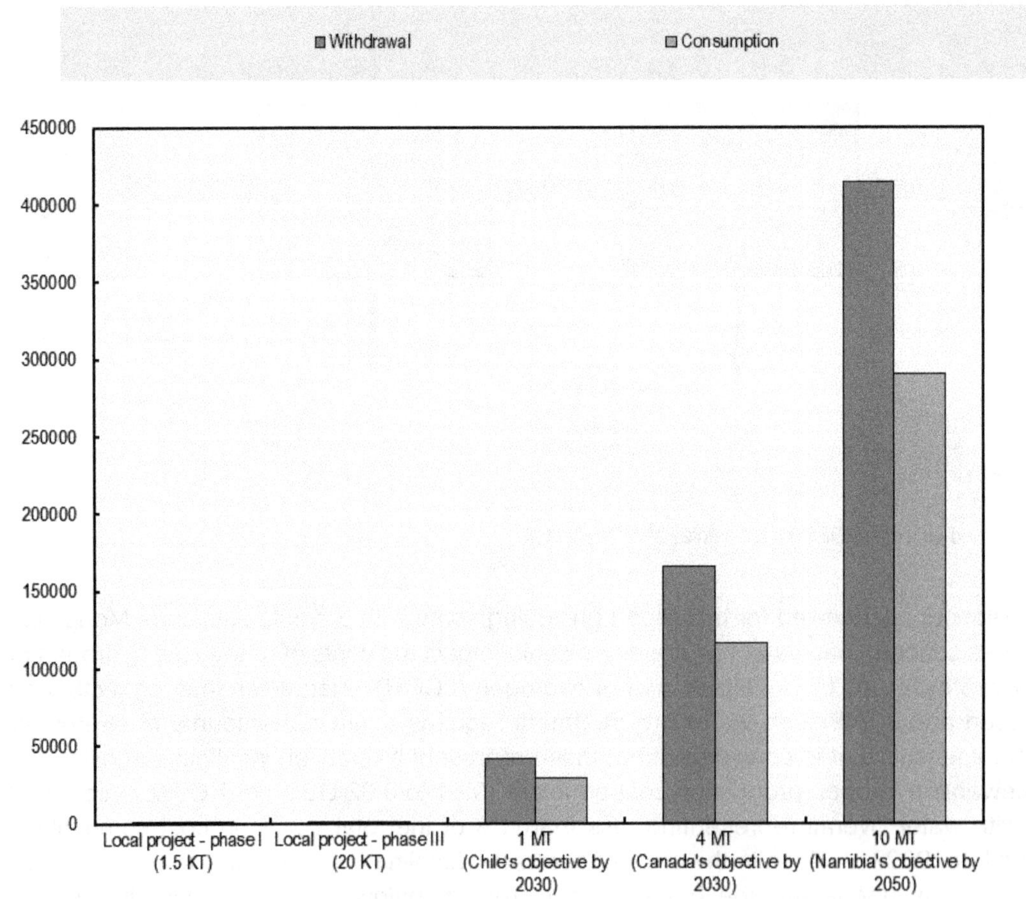

Source: OECD calculations

The exercise shows that while water needs for current renewable hydrogen projects under development appear relatively limited (below 600,000 m^3 of consumed water per year for 60 kilotons (KT) annual production of hydrogen), developing the sector at the same scale as other countries positioning themselves as competitive hydrogen exporters brings these needs to another scale (around 30 million m^3 of consumed water per year for a 1 MT sector). Water withdrawal figures for renewable hydrogen projects appear relatively small, but additional water use in an already water-stressed region such as the South Gobi Desert calls for caution. Impacts should be assessed carefully, while also considering the economic benefits of Mongolia developing a frontier green technology through pilot projects, as discussed in the other chapters of this report.

Renewable hydrogen is the focus of this report. However, some stakeholders in Mongolia are also exploring alternative hydrogen production methods, including coal gasification with CCUS. It is important to note that the water withdrawal intensity of renewable hydrogen is the lowest of all hydrogen production pathways (IRENA and Bluerisk, 2023[6]). In the case of coal gasification, additional water requirements can be necessary for cooling. Additionally, coal typically undergoes dewatering before gasification, creating significant additional wastewater management requirements. Notably, one of the hydrogen pilot projects in Mongolia is building on a previously abandoned coal gasification initiative (Guo et al., 2020[10]). According to IRENA estimates, producing hydrogen through coal gasification with CCUS consumes more than twice as much water as producing hydrogen through electrolysis. Without CCUS, coal gasification's water consumption is still 39% higher than that of renewable hydrogen production[3]. Not only is the electrolyser production pathway the most aligned with low-carbon targets, but it also supports more efficient water resource management objectives.

The development of a renewable hydrogen sector in Mongolia risks compounding the existing water stress in the South Gobi region

Despite concerns regarding the "water problem" of renewable hydrogen, the available literature concludes that, on a global scale, the water needed for renewable hydrogen production to reach net-zero objectives will only be a fraction of the water volumes currently consumed annually worldwide. The Tonelli et al. study found that water demand for renewable hydrogen production, even under worst case scenarios, would be less than 3% of the global combined water withdrawals for agriculture, industry, and municipalities, which currently stands at about 4,000 billion litres in total. Other studies consider that hydrogen needs would represent less than 1% of total water demand.

While these studies take a global perspective, the water problem associated with renewable hydrogen needs to be considered at the local level rather than globally. It will be important to assess both current and future water availability needs, as increased demand can aggravate water stress problems in locations that are already affected. A great number of emerging economies positioning themselves to become leading renewable hydrogen exporters, including countries in South America (Chile), the southern region of the African continent (South Africa, Namibia), and the Middle East and North Africa region (Oman, Qatar, Egypt, Morocco), are already exposed to water stress, similar to Mongolia (WRI Aqueduct, 2024[11]). These locations possess great potential for renewable energy production at competitive costs, particularly due to their abundant solar resources. However, being among the sunniest locations, these also tend to be among the driest. While most of these countries plan to leverage their access to seawater and use it in a desalinated form for renewable hydrogen production – a solution that comes with its own economic and environmental challenges - Mongolia lacks this option and must explore alternative approaches to address the water scarcity issue.

Box 4.1. Water withdrawal, water consumption and water stress

Water withdrawal, or water abstractions, is defined as freshwater taken from ground or surface water sources, either permanently or temporarily, and conveyed to a place of use. This measure helps to understand water demand from industrial, agricultural, and domestic users.

Water consumption is the portion of water use that is not returned to the original water source after being withdrawn. Consumption occurs when water is lost into the atmosphere through evaporation or incorporated into a product or plant and is no longer available for reuse.

Water stress is the ratio of water demand to renewable supply and measures the competition over local water resources. A ratio over 40% indicates high water stress, while a ratio over 80% indicates extreme water stress.

Pressures on freshwater resources are mounting: the OECD projects that by 2050, global water demand will rise by 55%, and 40% of the world's population will likely be living in severely water-stressed river basins.

Among other things, freshwater availability is affected by water abstractions (taking water from lakes, rivers, and underground sources), with over-abstraction leading to low river flows, depleted groundwater, desertification, biodiversity losses and potential social health issues. Well-designed allocation regimes are needed across sectors to ensure water is allocated where it can create the most value economically, socially, and environmentally.

Source: (OECD, 2024[12]), World Resources Institute

Mongolia, much like the countries of Central Asia, is affected by water stress, particularly in the region of Ulaanbaatar and in the south of the country. In theory there is no water stress issue at the national level, as total water withdrawals only represent 1.5% of total renewable resources. Spatial differences, however, result in local water stress, affecting local ecosystems (ADB, 2020[13]). The problem notably stems from the specific characteristics of Mongolia, a vast country, sparsely populated, with harsh climatic conditions leading to inaccessible surface water for long periods of the year and additional costs to infrastructure building and operation (Anna Nilsson, 2021[14]). The lack of quality water infrastructure is a serious impediment to access to water and concerns all sectors of the economy. According to the UNICEF, only 30% of households have access to safe drinking water and 56% to proper sanitation facilities. Moreover, 38% of groundwater wells do not meet the national standards for drinking water (UNICEF, 2023[15]). In all scenarios, water demand is expected to exceed current water supply by 2040 or even sooner (OECD, forthcoming).

Mongolian renewable hydrogen project developers consider the South Gobi region as the most promising location for renewable hydrogen development. This region possesses the highest potential for combined wind and solar renewable energy production, which is considered the most cost-competitive form of renewable energy production (Anna Nilsson, 2021[14]). With China identified as a promising export market by Mongolian investors, the South Gobi region's proximity to China's Inner Mongolia region presents a strategic advantage. The region also offers access to potential demand in the domestic mining sector. All current renewable hydrogen projects under development are located in this region. At the same time, while the region has abundant renewable energy generation potential, this generation infrastructure does not yet exist. As discussed in Chapter 3, the development of this infrastructure is essential to establish a renewable hydrogen industry in the area. Ordinarily, access to the grid would play an important role in determining project location, enabling hydrogen production site to draw on power from the grid when necessary and provide surplus electricity back to the grid.

At the same time, the South Gobi Desert is also the region where water stress is the most acute, together with the Ulaanbaatar region. This stress is exacerbated by the planned development of the mining industry, which is set to put additional pressure on local water resources. In this region, characterised by no surface runoff and very little rainfall, reliance is placed on non-replenishable groundwater sources. These sources remain under extreme pressure and are rapidly depleting, a phenomenon accentuated by climate change. Furthermore, a lack of data prevents knowing exactly what the capacity of these aquifers is and whether some of them are replenishable. Surface water, on the other hand, is concentrated in the northern part of the country. The spatial mismatch between renewable energy potential and water availability is not uncommon, as mentioned above, since regions with favourable sun and wind conditions often face water stress problems (Tonelli et al., 2023[5]).

Given the geo-spatial mismatch between the country's abundant surface water resources in the north and its highest combined solar and wind generation potential in the south, it may be advisable for the government to gain a clearer understanding of the economics and environmental externalities of intraregional electricity transfers. These transfers could facilitate renewable hydrogen production in the north, where ground water is accessible, versus utilising aquifers in the south. This would imply the need to build additional hydrogen pipeline infrastructure to transport the hydrogen produced in the north to domestic and external off-takers in the south. What is certainly clear is that the cheapest way to deliver hydrogen to end-users is to cluster and co-locate the production infrastructure with industrial demand. The greater the distance between them, the higher the associated costs. In addition, under the current assumptions that the value proposition of Mongolia as a potential hydrogen producer is the quality of its combined wind and solar power, private sector projects are more likely to be drawn to the south of the country. In this region, they are not only close to major domestic off-takers but also to export markets in northern China. As discussed in Chapter 3, given the substantial costs involved in building the pipeline infrastructure for export, it is uncertain that the economics of Mongolian hydrogen exports would remain competitive if there was a need for additional pipeline infrastructure to be built to transport hydrogen produced in the north, where production with combined wind and solar is likely to be less efficient, to the south.

There is a serious risk that the development of renewable hydrogen in Mongolia could encounter challenges. These risks arise from either a lack of access to water, if authorities do not allow water abstraction for hydrogen projects, or from the creation of a new competition for water usage at the local level, if such authorisation is granted without a clear plan for managing this new demand. Both cases would be detrimental to the development of the sector. A lack of water access, compounded by insufficient data about water availability, will discourage investment. Moreover, competition with other users may create local tensions and lead to delays in project implementation or generate discontent and lack of support from the population, which may also deter investment. Given recent tensions around water access between mining operators and herders, a more thorough and transparent assessment of water availability in the region could significantly politically de-risk future projects[4]. From the perspective of investors, uncertainty regarding water access for renewable hydrogen production is a major risk. While the water volumes needed for pilots and demonstration plants are relatively small, the problem can become acute when considering the development of giga projects, or the scale-up of local production that could be needed to reach cost-competitiveness.

Public measures with potential for being "game changers" are discussed in section 4.4 of this chapter. These measures include water allocation regimes and plans that are aligned with future social and economic development strategies. For example, if the government is expected to articulate a plan to phase out the production of coal, this would imply declining demand from that activity on the aquifers in question, consequently affecting future water availability for other activities including renewable hydrogen production. Conducting a socio-economic and environmental cost-benefit analysis of continued aquifer use for coal mining compared to hydrogen production could help to inform public policy and social discussion of these issues. Another challenge for the government is to rationalise who should be responsible for improving water data in the South Gobi. Additionally, in the context of its ambitions to develop the mining sector or hydrogen production, there is a role for the public sector to assume the responsibility of clarifying water availability in the region.

Renewable hydrogen project developers in Mongolia expect to be able to access local groundwater. However, as of the time of this report, no formal agreement had been reached with local authorities, including concerning environmental impact, based on the information collected. Developers hope to be granted access to local groundwater sources on the grounds that water demand for renewable hydrogen is considerably smaller than water demand of the mining sector (Fig. 4.4.). From the perspective of public policy, it is crucial to extend the approach beyond such consideration and encompass all aspects of a sound and robust water allocation regime. This means considering all available water supplies and both present and future demands, and then deciding on an allocation that is fair and aligned with criteria for maximising economic benefit. The principles of a water allocation regime that is aligned with OECD standards and good practices are discussed in Section 4.4 of this chapter.

When considering the water demand and use in energy and industrial activities, it is important to account for another parameter, the potential for wastewater recycling. This factor directly depends on water recovery rates, the quality of the water recovered, and the existence of infrastructure supporting wastewater collection, treatment, and re-use. Typically, recovery rates in the renewable hydrogen sector are expected to be much lower than in the mining sector, which benefits from extensive experience in the optimisation of water use, including through recycling. However, improvements in water technology and management will likely develop as the renewable hydrogen technology matures. Estimates presented in Figure 4.1 consider a recovery rate of 30% for renewable hydrogen production, whereas recovery rates in the mining sector are considerably higher. Oyu Tolgoi, for example, boasts a water recycling rate of 85-87%[5].

Comparing water consumption by the renewable hydrogen sector relative to other water uses in Mongolia can provide valuable insights for discussing water availability for production purposes. Figure 4.3 below represents water demand by user and provides a view of water demand volumes for renewable hydrogen production. A renewable hydrogen sector with an annual capacity of 1 MT would require similar water resources comparable to the projected needs of the mining sector in the South Gobi mining and heavy industry region in 2030, considering consumption levels. As Mongolia deliberates on water allocation in the South Gobi region and explores the possibility of engaging in consultations to develop a comprehensive hydrogen strategy and establish export production targets, this comparison becomes an important consideration.

Figure 4.3. Water demand of a potential renewable hydrogen sector relative to existing water users in Mongolia

In thousand cubic meters

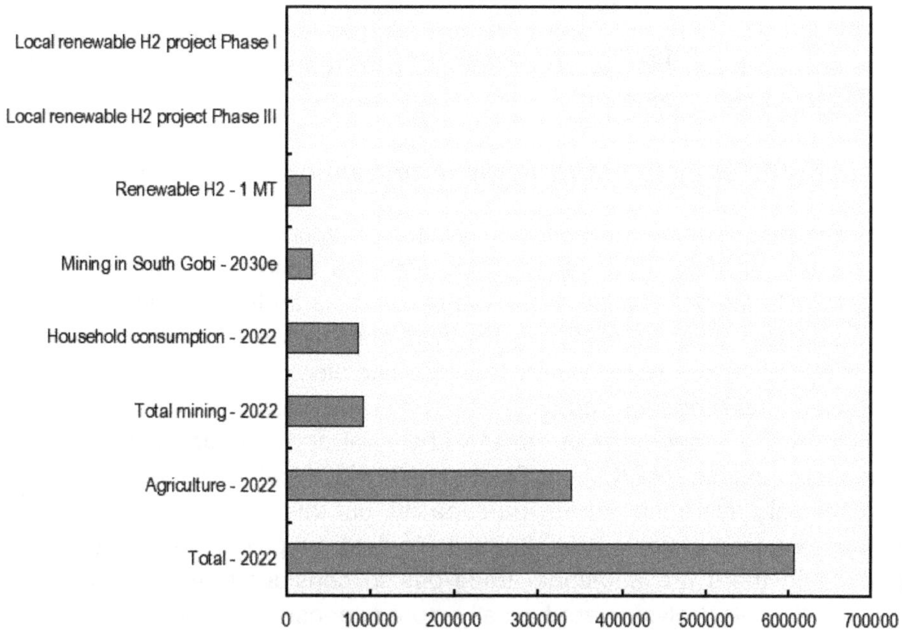

Note: 2030 water demand of the mining sector in the South Gobi region was estimated in a 2021 study by the 2030 Water Resources Group. The estimate in this chart corresponds to a medium scenario (the study also envisages a low and a high scenario).
Source: OECD, Water Agency of Mongolia, 2030 Water Resources Group

The way that water is used and managed in renewable hydrogen production can, in turn, lead to risks related to biodiversity. Groundwater abstraction related to new hydrogen operations can directly affect groundwater-dependent biodiversity. The disposal of brine resulting from groundwater desalination needs proper management, as brine is highly saline and can contain treatment chemicals that may affect local ecosystems where the brine is discharged. Although the brine resulting from groundwater pre-treatment may contain less salts than brine discharged as a result of seawater desalination, the former cannot be returned to the sea for dilution, and at this stage it is not clear how pilot projects intend to manage wastewater and brine (e.g., through evaporation ponds or discharge into the local environment). These risks need to be assessed and effectively mitigated, ideally within a broader framework that integrates biodiversity into infrastructure development. Mongolia has a number of tools in place that can be leveraged to develop a thorough approach to delivering biodiversity-aligned infrastructure projects. These include Environmental Impact Assessments (EIA), which are already required by national water regulations as discussed in section 4.4. Additionally, a range of regulatory, economic or voluntary instruments can be mobilised to catalyse the integration of biodiversity considerations into renewable energy and hydrogen infrastructure projects (OECD, 2024[16]).

Available practices and solutions for reducing pressure on freshwater sources for renewable hydrogen production

Anticipated local water stress caused by a growing hydrogen production is prompting experts, policymakers and industry stakeholders to consider sector-specific solutions to addressing the challenge.

Beyond the selection of seawater desalination, which is not suitable in Mongolia's case, there are other promising avenues to explore. Some of these avenues are sector-specific and seek to reduce the water footprint of renewable hydrogen production. While they may not singularly resolve existing water stress in the South Gobi Desert, they offer incremental efficiencies and could foster local innovation over time. These solutions require that policymakers identify and implement appropriate and incentives to promote their adoption. This section looks into two complementary sets of solutions, and how governments can encourage their adoption. The first set focuses on promoting water-efficient technology development and adoption, while the second highlights the opportunity to use treated wastewater as an alternative water source for renewable hydrogen production.

Encouraging water efficiency in technological innovation and industry development

Technological choices can determine water efficiency in the production of renewable hydrogen at all stages of the industrial process, as they do in other water-intensive economic activities, such as agriculture (irrigation), light manufacturing and mining. For example, studies by the International Renewable Energy Agency (IRENA) found that when considering the two main types of electrolysis technologies, PEM electrolysers tend to consume less water overall than alkaline electrolysers on average (-22%). Moreover, the efficiency of the electrolyser directly impacts water requirements, with a one percentage point increase in efficiency resulting in a 2% decrease in water withdrawal and consumption, primarily due to water savings during the cooling phase (IRENA and Bluerisk, 2023[6]). Alkaline electrolysers currently represent 60% of installed renewable hydrogen production capacity, but this is expected to change in the coming years, according to announced projects (IEA, 2023[1]). To date, no project developers have finalised their technology choices, and there are additional trade-offs to consider from an economic and technical standpoint. Currently, PEM electrolysers are typically more expensive and require more energy to operate than alkaline electrolysers, although costs are reducing rather rapidly. There is no clear information about the water consumption of other maturing electrolyser technologies, including the solid oxide cell (SOEC), which currently represents less than 1% of the total installed capacity, and anion exchange membrane (AEM) which is a technology under development (IEA, 2023[17]). Cooling technologies vary in their water intensity. The two conventional cooling methods are dry and wet cooling. Dry cooling is more water-efficient yet tends to be more costly, energy-intensive, and less suitable in warm climates. Regardless of the method chosen, water efficiency needs to be assessed end-to-end.

Renewable hydrogen is a dynamic field in terms of research and development (R&D), with ongoing research focuses on technological improvements aimed at enhancing water efficiency. Efforts are underway to find solutions that would allow electrolysers to be fed with low-grade, saline water and even seawater, thereby reducing freshwater needs and water treatment costs associated with renewable hydrogen production (Tong et al., 2020[18]) (Marin et al., 2023[19]). Technology improvements in cooling systems can also yield better water outcomes for renewable hydrogen production installations. Researching air-cooling technologies for electrolysers, for example, can help reduce freshwater consumption (IRENA and Bluerisk, 2023[6]). Additionally, research is exploring technologies for recovering water from the consumption of hydrogen itself. When hydrogen is consumed, it releases pure water that returns to the environment through evaporation. Some researchers are looking into methods to capture this water as it is being released, aiming to extract value from it. Since the place of recovery would in many cases be different than the place of water consumption for hydrogen production, the potential of such technology to reduce water stress locally would be limited. It could, however, help to put a price on this water and value it in the commercialisation process.

As the Mongolian government devises its national approach to renewable hydrogen development, policies and measures in the support of technology adoption and project implementation can integrate criteria relating to water efficiency. For example, as Mongolia seeks to promote Foreign Direct Investment (FDI) and foster linkage with Small and Medium-sized Enterprises (SMEs) to enable technology transfer, targeted efforts could focus on attracting Multinational enterprises (MNEs) with expertise in freshwater solutions suitable for Mongolia's context. In mobilising domestic research and innovation resources for renewable hydrogen development, there could be an opportunity to specifically target water efficiency as a priority, by establishing a specific cross-sectoral working group focusing on water-hydrogen optimisation. In the longer run, as projects and renewable hydrogen sectors develop, the government could partner with industry stakeholders to monitor water usage data and benchmarks, both domestically and internationally, and work toward establishing industry standards and promoting best practices adoption. Stakeholders could also assess the relevance and opportunity to bring water efficiency knowledge from the domestic mining sector, which faces significant water scarcity challenges but has accumulated decades of experience and achieved high rates of water recycling.

While policies supporting the integration of water efficiency in technology development cannot singularly ensure sustainable water management in Mongolia's context of renewable hydrogen development, they can serve as a strong signal to direct R&D and innovation investments towards solutions compatible with sustainable water use across industries. Ideally, this approach should be implemented across all uses, including irrigation, mining, and light industry. Renewable hydrogen can serve as a pilot for developing such an approach. Given the country's water-related constraints, it offers a chance for Mongolia to establish a framework conducive to water technology transfer within the domestic economy. Moreover, under the right conditions, it could position Mongolia to become an originator of water-related innovation.

Mongolia can investigate initiatives promoting treated wastewater use for renewable hydrogen production and developing circular economy approaches

Excluding seawater desalination, using treated wastewater as feedstock for renewable hydrogen production is emerging as the key solution for reducing pressure on freshwater resources and mitigating competition with water needs for agriculture and human consumption. Some OECD countries, such as Portugal and Australia, have a clear objective of integrating wastewater and renewable hydrogen sectors under their national strategies. Pilot projects are being developed in these countries and in others, including Spain and the United Sates. A significant body of research is focusing on this production pathway, with examples including research conducted at the Monash University in Australia and within the EU (Maddaloni et al., 2023[20]).

Government and private initiatives are promoting the use of wastewater as feedstock for the production of renewable hydrogen. Portugal's national hydrogen strategy explicitly refers to the use of urban and industrial wastewater as feedstock for renewable hydrogen production. In Australia, several studies and pilot projects are underway to explore the opportunity of co-locating wastewater treatment plants and hydrogen facilities (Water Services Association of Australia, 2021[21]). In Spain, Cepsa, one of the world's largest energy companies, has signed an agreement with the public water company ARCGISA to utilise urban wastewater effluent as a feedstock for the renewable hydrogen facilities it is building in Andalusia (Box 4.2). Apart from alleviating pressure on freshwater resources, the utilisation of urban or industrial wastewater for renewable hydrogen production is viewed as an opportunity to enhance synergies between the energy and water sectors, creating a new avenue for investment and incorporating circular economy principles by assigning economic value to an underutilised resource. Consequently, public and private organisations from the water sector are taking an interest in the development of renewable hydrogen, with numerous examples of water utilities or technology companies forming partnerships and consortia to advance renewable hydrogen projects.

Box 4.2. Producing renewable hydrogen from wastewater effluent in the Andalusian Green Valley

The Andalusian Green Hydrogen Valley in Spain is a project developed by the energy company Cepsa, targeting a combined capacity of 2 GW of electrolysis at two centers located in Palos de la Frontera in Huelva and San Roque in Cádiz, for a total annual production of 300,000 tons of renewable hydrogen by 2030.

In December 2022, Cepsa announced that it had reached an agreement with the public company Aguas y Servicios del Campo de Gibraltar (ARCGISA) to supply its renewable hydrogen operations at San Roque Energy Park with treated urban wastewater. Water from urban runoffs of Los Barrios and San Roque will be treated and purified in a new wastewater treatment plant to be built near the park. With this project, 4.2 million cubic meters of urban wastewater will be reused instead of being discharged in the sea.

On the 31 January 2024, Cepsa's San Roque Energy Park already inaugurated a new wastewater reuse plant, which enables a 25% reduction in water use in industrial operations. The water treated at this plant will be reused in Cepsa's power plant's cooling towers. The project required an initial investment of 2.5 million euros.

These projects are part of Cepsa's environmental commitment and transformation plan, which entail the target of reducing freshwater capture in water-stressed areas by 20% by 2025, compared to its use in 2019.

Source: Cepsa

Wastewater use and circularity can be considered as part of Mongolia's future renewable hydrogen strategy. Urban runoff from cities and towns could be considered for the production of renewable hydrogen, noting that Mongolia's wastewater treatment plants are outdated, and that this infrastructure needs investments (Asian Development Bank, 2020[22]). In the city of Ulaanbaatar, for example, the daily flow of wastewater is 200,000 m3/day, projected to reach 580,000m3/day by 2030. Effluent water is released in the environment and is not reused. USA's Millenium Challenge Corporation (MCC) is currently working on a project to build a water recycling plant that would reuse 50,000m3/day of treated urban wastewater from Ulaanbaatar to support the operations of two power plants located near the city. An additional 200,000m3/day flow of treated wastewater would be released in the Tuul river. One option worth exploring is the potential use of this water for renewable hydrogen production, although a clear cost-benefit analysis of this option has yet to be developed. The idea of co-locating wastewater treatment facilities with industrial activities that have a demand for hydrogen and renewable energy, as well as hydrogen production could be particularly relevant in areas with sizeable mining operations in the South Gobi Desert. As stakeholders consult on the development of a comprehensive renewable hydrogen strategy, they could also assess the feasibility of developing a pilot project to test a circular approach. This approach would involve collaboration between the water sector, the mining industry and the energy and hydrogen sectors.

Using wastewater effluent as a feedstock for renewable hydrogen production will require some questions to be addressed related to the viability of such projects, both economically and environmentally. From an economic perspective, projects aiming to use wastewater as feedstock for renewable hydrogen production will need to overcome the challenges stemming from the current lack of information and management of this wastewater. Under current circumstances, wastewater is a by-product that provides no economic value. Using wastewater as a feedstock for renewable hydrogen could potentially assign value to wastewater and provide a rationale for improving its management. However, achieving this would require the development of the necessary capacities. Anecdotal evidence from outside of Mongolia suggests that wastewater facility management often lacks the framework to determine a selling price for wastewater. Moreover, the business model should incorporate the costs for transferring the wastewater from its original location (the Tuul river) to its destination where it will be used (the South Gobi). From an environmental perspective, the impact of such projects on the environment should be assessed, as wastewater effluents can benefit receiving ecosystems by increasing water flows and carrying nutrients and other materials. Consequently, these projects would have to undergo EIAs, as required and detailed in Mongolian law (Government of Mongolia, 2023[23]). These assessments are a critical tool for evaluating such impacts and documenting the consequences of the wastewater effluent removal on the environment.

In its investigation, Mongolia could look at the example of Israel, a country that implemented policy reforms to promote wastewater reuse. These reforms included transferring responsibility for water treatment from municipalities to municipal and regional water companies (OECD, 2015[24]). As a result, 85% of wastewater in Israel is treated and reused, primarily for agricultural purposes, but also for the industrial use. The reuse of wastewater has helped Israel address inter-annual and inter-seasonal variability and increase resilience in the face of climate change (OECD, 2015[24]).

More broadly, stakeholders are exploring the concept of co-locating key infrastructure – such as industry, transport, energy, water, and waste - to establish economic hubs based on shared assets and promote circularity. These hubs can take various forms, including clusters, industrial parks, special economic zones, or hubs, and may operate under special regulatory and institutional frameworks. They are typically utilised as tools for attracting investments in specific locations and sectors, aiming to build economies of scale and benefit from agglomeration economies. The sharing of indivisible facilities, such as public goods or infrastructure serving several individual of firms, is one the mechanism through which agglomeration economies are produced (Duranton and Puga, 2004[25]). These models also bear a potential for cost savings through shared infrastructure, reduced transportation costs due to production being situated near demand, and opportunity to put an economic value on excess, co- and by-products of industrial processes. For example, the oxygen produced as a co-product of hydrogen production can be used for as an input to water treatment processes (Water Services Association of Australia, 2021[21]). Such hydrogen and industrial hubs have the potential to promote sustainable diversification at the sub-national level. A recent OECD report on mining regions and cities, suggests that Antofagasta, Chile, a copper-producing region facing increasing water stress, stands to benefit from its significant potential for renewable energy and renewable hydrogen production. Through the strategic utilisation of synergies and the implementation of an integrated approach between renewable energies, net-zero policies and water management, Antofagasta could effectively tackle its challenges (OECD, 2023[26]). Consideration of similar synergies could be envisioned for the South Gobi region in Mongolia. The development of a comprehensive strategy would provide an appropriate platform to test this idea and formulate the planning and incentives necessary to support such a model. This approach holds the potential to facilitate sustainable diversification and a just transition.

Strengthening Mongolia's water institutional and policy framework to enable a cross-sectoral approach of water and hydrogen

Governments' role is pivotal in managing the competing uses and constraints of the water sector. Water is fundamental to human lives, a valuable resource for economic activities, and at the centre of entire ecosystems. Given the magnitude of water needs of the hydrogen sector, should this sector develop in Mongolia, the strength of the country's broader water institutional and policy framework will be pivotal. It will play a key role in managing the emerging competition with other water needs, ensuring that water access is not a key obstacle to investment in the sector, and enable the emergence of solutions discussed in the second section of this chapter. Acknowledging water-related risks in the country and weaknesses in its water framework, the Mongolian government has been adopting policies and strategies in the past 15 years to improve and align with international good practices, with successful results. In addition, the government is actively exploring a water infrastructure plan that could help address the spatial mismatch between water location and demand, particularly focusing on alleviating the water scarcity issues affecting the Southern region. Mongolia's success in implementing water governance and policy reforms, bridging existing gaps and moving forward its water plans will contribute to confirming whether and how renewable hydrogen production can access the necessary water sustainably while also providing economic value to the country.

A cross-sectoral approach to water management requires strong co-ordination, enhanced capabilities at sub-national level, and strengthened data management

Mongolia has been reforming its water sector to enhance its governance at national and subnational levels, improve data collection and strengthen its water management system. Mongolia's main strategic water policies are contained in the National Water Program (2010), National Security Concept (2010), Integrated Water Management Plan (IWMP) (2013), Green Development Policy and Mongolia Sustainable Development Vision 2030 (2016). The laws regulating access to water and usage are the Mongolian Water Law (revised in 2012), Natural Resources Use Fees Law, and Water Polluting Fees Law (updated in 2019). In addition, Mongolia's long-term economic development plan, Vision 2050, adopted in 2020, incorporates comprehensive strategic water objectives, further reinforced in the country's action plan, the New Revival Policy adopted by the Mongolian Parliament in 2021. Within this policy framework, the government highlights the 'Blue Horse' water project (the Blue Horse project), designed to improve water access across the country, including in the South Gobi region, as discussed later in this chapter.

The Mongolian government has clearly identified the improvement of the governance and management of its water sector as a priority for sustainable economic growth. While strategic plans and documents outline measures that could contribute to more efficient allocation and utilisation of water resources, along with other environmental benefits, implementation remains a significant hurdle. Chapter six of the 2050 Vision on Green Development includes objective 6.3 to prevent water scarcity, preserve surface water and enhance framework conditions to meet the water needs of the Mongolian society. Several measures are identified under this objective and phased in three phases (Phase I - 2021-2030, Phase II - 2031-2040 and Phase III - 2041-2050). During the first phase, the government plans to lay the ground for a tiered water tariff system with the aim to raise water prices to incentivise water efficiency and promote reuse. Additionally, efforts will be made to improve compliance with regulations surrounding defined water reservoir protection and water supply zones, create water reservoirs and enhance water data through resource exploration and mapping. Measures to reduce water pollution and address shortages are also outlined. Phases II and III include the continuation of efforts initiated in Phase I, with a focus on refining a pricing system and introducing new water-saving technologies. Despite these outlined plans, local stakeholders note a lag in the implementation of water policies, with anticipated benefits yet to materialise. This gap is attributed to a lack of political commitment, resulting in the absence of robust implementation plans and insufficient financial and human resources. The second part of this assessment at least is substantiated by several reports covering the water sector of Mongolia, and the discussion below highlights the main challenges and possible remedies.

The current water institutional framework in Mongolia is relatively complex and requires further and better co-ordination across actors (Fig. 4.5). At the centre of government, 10 ministries participate in the water sector, including the Ministry of Environment and Climate Change (integrated water policy, protection, increase and improvement of water resource use, water inspection and monitoring), Ministry of Foreign Affairs (international water co-operation and agreements), Ministry of Food, Agriculture and Light Industry (irrigation, water supply for food and light industry), Ministry of Energy, Ministry of Finance and Ministry of Economy and Development. A National Water Council operating under the Prime Minister was created in 2016 to oversee cross-sectoral coordination. Nonetheless, and this is a common challenge for cross-sectoral councils dealing with complex issues such as water management, its ability to gain traction, in view of competing priorities at the Prime Minister's level, is uncertain according to local stakeholders.

Figure 4.4. Water institutional framework of Mongolia

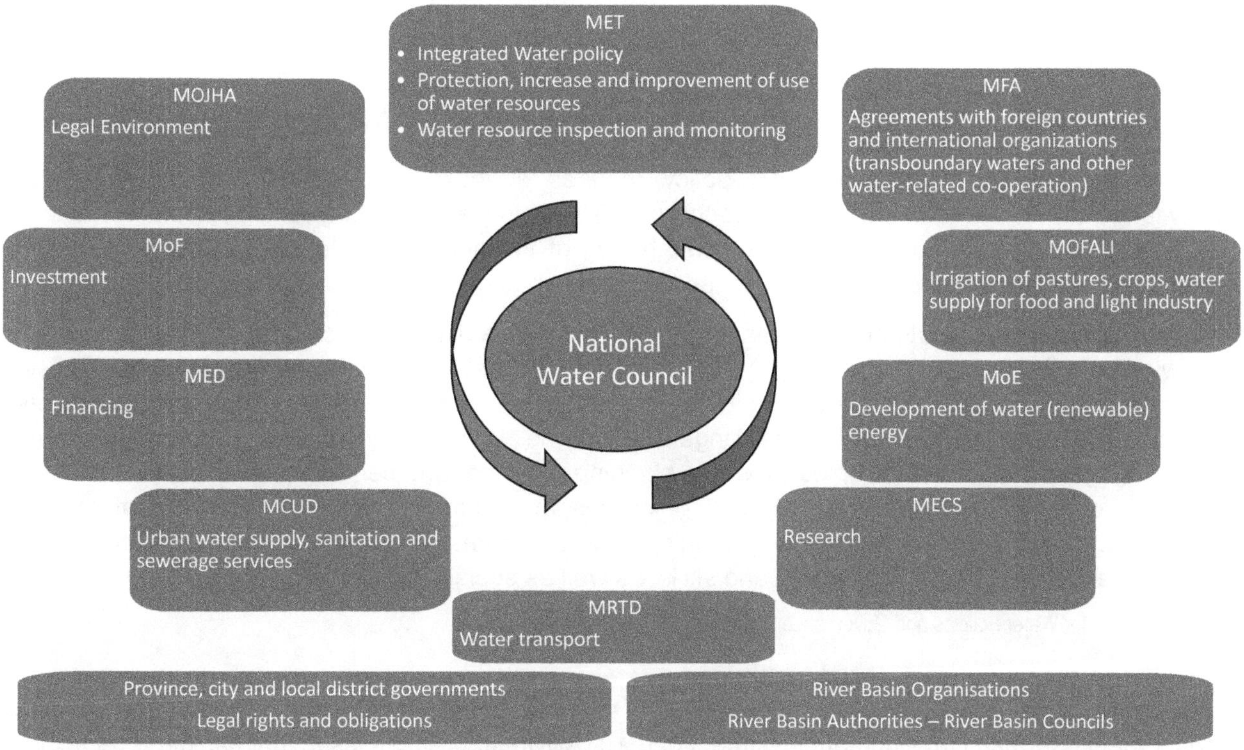

Note: MECC: Ministry of Environment and Climate Change (formerly Ministry of Environment and Tourism); MFA: Ministry of Foreign Affairs; MOFALI: Ministry of Food, Agriculture and Light Industry; MoE: Ministry of Energy; MECS: Ministry of Education, Culture and Science; MRTD: Ministry of Road and Transport Development; MCUD: Ministry of Construction and Urban Development; MED: Ministry of Economy and Development; MoF: Ministry of Finance; MOJHA: Ministry of Justice and Home Affairs.
Source: Author

In recent years, Mongolia established 21 River Basin Organizations (RBOs) for its 29 water basins as a means of decentralising water management in the country. This initiative aligns with international best practices, as noted by the OECD (OECD, 2021[27]). Eventually, 29 RBOs should be established (ADB, 2020[13]). These RBOs are responsible for the implementation of the plans contained in the Integrated Water Resources Management Plan (IWPM) through multisector stakeholder consultations. Each RBO comprises one River Basin Authority (RBA), composed of 5 to 12 government personnel reporting to the Ministry of Environment and Climate Change, and one River Basin Council (RBC), composed of 35 representatives of the government, water users and other stakeholders from the civil society and academia. RBAs are responsible for managing water resources, setting water abstraction limits ("caps"), and issuing permits. As water users, renewable hydrogen operators need to establish a water use contract with RBAs and submit water use reports. RBCs aim to involve local communities, offering them an opportunity to review and comment on both local water plans and the work conducted by RBAs.

Local experts and international partners' reports highlight the need to strengthen water governance in Mongolia and enhance human capital and skills within water institutions. In particular, strengthening the capabilities of the recently created RBOs and clarifying their roles will support better water governance and management at the sub-national level. Given that RBOs in Mongolia are relatively new entities, further training and capacity-building are important to fulfil their mandates. Drawing from the experiences of OECD

Council Recommendation on Water adherents, various programs and tools are available for this purpose. The Recommendation invites its Adherents to "adapt the level of capacity of responsible authorities to the complexity of water challenges to be met, and to the set of competencies required to carry out their duties" as a precondition to effective water governance (Box 4.3) (OECD, 2016[28]). These insights provide valuable guidance for enhancing the capabilities of Mongolia's RBOs and advancing water governance efforts.

Box 4.3. Adapting the level of capacity to the complexity of water challenges at subnational level

1. Effective governance of water policy requires adequate skills, technical expertise, and knowledge as well as human and time resources. This can be particularly crucial in the context of decentralisation and ensuring the proper capabilities at the subnational level. A number of OECD members have devised tools and approaches to tackle capacity gaps and issues.

2. In **Ireland**, a Catchment Management and Science Unit was created to strengthen the knowledge base for river basin management and help target resources appropriately. In 2017 in **Australia**, the Council of Australian Governments published training modules of the National Water Initiative (NWI) on 'Considering climate change and extreme events in water planning and management' and 'Engaging Indigenous peoples in water planning and management'. Reviews of the NWI are required to assess progress against NWI objectives and commitments. In **Austria**, professional associations promote the education and training of water professionals and institutions, focusing on "soft" capacities such as management and skills, as well as social expertise.

Source: Toolkit for Water Policies and Governance, OECD, 2021

Assessments also highlight the opportunity to enhance multi-stakeholder involvement in water management (German Development Institute, 2020[29]) (ADB, 2020[13]), an issue that has been particularly critical in the framework of mining operations. These assessments find that RBCs have a limited role as multi-stakeholder platforms. They often suffer from overrepresentation of lower-level government officials, and face challenges due to insufficient funding for organisation meetings, particularly for covering the expense of stakeholder travels over long distances, necessitated by the geographical coverage of these bodies. EIAs, which are mandatory in the process to obtain a water permits for use exceeding 100 m^3 per day, require to lead public consultations, which is another mechanism for involving local communities. Assessments, however, find inconsistencies in leading such consultations and utilising their outcomes as input for the final decision-making, which is undertaken at the Ministry of Environment level. Reports from the German Development Institute and Asian Development Bank (ADB) point out the absence of records of such consultations, alongside general mistrust and lack of transparency, given that the granting of water use licenses does not undergo public scrutiny. Clearer guidelines and transparency in these processes could strengthen governance in the water sector, enhance representativeness of stakeholder diversity and mitigate the risks of local conflict and resistance that may arise with the development of new water-using industries such as renewable hydrogen.

Improving water data systems would greatly help with enhancing Mongolia's water sector. Mongolia faces water data scarcity due to insufficient testing and monitoring at the local level and poor data management systems, hindering access to up-to-date data (German Development Institute, 2020[29]). The OECD highlights that monitoring groundwater use in particular is technically demanding and costly (OECD, 2017[30]), while local stakeholders acknowledge that knowledge of available groundwater data in Mongolia is limited and would require further exploration and hydrological studies. Additionally, access to groundwater data is hindered by data confidentiality regulations (2030 Water Resources Group, 2021[31]). The lack of quality and accessible water data is an impediment to sound, evidence-based water policy planning, implementation, and enforcement, posing obstacles for project developers needing water for their operations. It also impedes informed, open multi-stakeholder dialogues on water uses. The initiative led by the 2030 Water Resource Group, in collaboration with the Ministry of Environment and Climate Change of Mongolia, to create an interactive groundwater monitoring digital dashboard is a welcome step in this regard. Furthermore, Mongolia has recently been introducing Geographic Information Systems (GIS) and Remote Sensing tools which should also contribute to improving water data in the country. Through the National Dialogue on Water in Mongolia held in 2024, the OECD has been developing recommendations to further improve water information systems (Box 4.4).

Box 4.4. Water information systems

Hydrological observation stations are essential for collecting data used in water resource planning, development, and management. To support these functions, the government operates 154 surface water stations and 325 groundwater stations, including 68 automated, 255 semi-automated, and 2 manual stations. Of the 325 groundwater stations, 28% are located in Ulaanbaatar city, 45% in the Gobi region, and 27% in the rest of the country, with observations every six hours. Provincial meteorological and environmental stations monitoring surface water usually have one to six observers. Mongolia currently has one observer per 5,000 km^2, which is below the WHO recommendation of having one rainfall observation station per 250 km^2 in mountainous areas and 575 km^2 in plains and hilly areas. However, given Mongolia's vast surface area - 40% of which is desert - applying these standards is impractical.

The majority of the 21 river basin organisations are in need of enhanced hydrological observation as they are unable to provide systematic data analysis. This would require additional facilities, budget, and specialised personnel.

In 2024, the National Dialogue on Water in Mongolia discussed how policies and information systems can help to meet the growing demand for water in Mongolia. The Dialogue was hosted by the Mongolian Ministry of Environment and Climate Change in 2023-2024, facilitated by Asia Water Council and the OECD. The Dialogue yielded several recommendations to strengthen water information systems:

- **Modernisation of the hydrological monitoring system**: Establishing standards for innovation and institutional support in the installation and operation of a systematic hydrological monitoring network. Expanding ICT-based systems and modernising ageing infrastructure could enhance data quality and flood forecasting and warning.
- **Integrated water information systems**: Developing a national water information system and water management centre to centralise data and streamline water management efforts.
- **Capacity building**: Creating a comprehensive training approach for water experts, including a mid- to long-term roadmap for skill development. A National Water Environment Education Centre could serve as a hub for ongoing training and education.
- **Pilot projects**: two pilot projects are proposed. The first proposed project is to establish an Integrated Water Resources Management (IWRM) Operations Centre within the Ministry of Environment and Climate Change. The second project is to develop a training curriculum on water information through the existing training centre of the Water Services Regulatory Commission (WSRC).

Source: A contribution by Asia Water Council, based on the Forthcoming report on the National Dialogue on Water in Mongolia (OECD and AWC, *forthcoming*)

These different components – more cross-sectoral co-ordination, resources and capabilities aligned with mandates at sub-national level, and improved data management – will be crucial for the development of the water-hydrogen sector. Interviews suggest that operators in the mining sector have developed capabilities to autonomously manage local water issues, including conducting their own water exploration and testing activities. However, the renewable hydrogen sector faces a greater challenge because it is nascent and lacks the resources, knowledge and local expertise of the established mining sector. The development of a new economic activity such as renewable hydrogen, where commercial risks are high as are expectations to comply with environmental and sustainability standards, require public support to de-risk any parameter affecting the sector's economic model as much as possible, and water clearly is one. In the framework of the development of a national strategy on renewable hydrogen, it is advisable for water authorities to participate in the creation of a specific cross-sectoral working group on the water-hydrogen nexus, alongside representatives from the Ministry of Energy and the industry. Such a working group could consolidate the available knowledge and information on water quality and availability in locations suitable for renewable hydrogen production, conduct impact analyses of water use for hydrogen production (including brine production), and, in turn, formulate recommendations to the industry for responsible water use and to the government for ensuring a regulatory framework that supports this sector's development while ensuring sustainability.

The legal and regulatory framework for water use in Mongolia can be further strengthened to ensure fair allocation across sectors, incentivise water efficiency and increase revenues

A sound legal framework for water use can facilitate access to water across users in a fair, sustainable, and transparent way, maximising economic and social benefits for the society. Water resources allocation regimes, defined as a combination of policies, laws and mechanisms that determine who is able to use water resources, how, when and where, are critical to manage the risk of shortage and adjudicate between competing uses (OECD, 2015[24]). The OECD's work acknowledges that existing water allocation regimes in OECD countries are not equipped to address today's challenges, which include intensifying competition across uses, climate change, and the objective of valuing water-related ecological services. As discussed earlier in this chapter, renewable hydrogen development can exacerbate local water use competition. In the context of Mongolia, where the mining sector is already putting pressure on scarce water resources in the southern region of the country, this risk is particularly acute. A robust legal framework for water usage can help mitigate the risks, ensuring water availability for renewable hydrogen production in case this becomes a priority. It can also incentivise efficient water use across user sectors.

The main laws regulating access to water for water users in Mongolia are the Mongolian Water Law (lastly revised in 2012), the Natural Resources Use Fees Law (2012), and the Water Polluting Fees Law (adopted in 2019). These laws regulate the conditions under which water users (industrial and service companies) and consumers (citizens and not-for-profit users including herders and farmers) can access and use water resources and discharge wastewater. Penalties for rule violations are also outlined in these laws. They contain a number of economic instruments with varying degrees of implementation and enforcement (2030 Water Resource Group, 2014[32]), including usage charges, wastewater charges and water pollution fee. Great steps have been made in strengthening and implementing these instruments, and this trend should continue (OECD and AWC, Forthcoming[33]). According to the Law on Water, renewable hydrogen project developers are considered to be industrial water users.

Water users must apply to government agencies for water use permits or licenses. Depending on the daily water volume requested, the application review will either be carried out by the Mongol Water State-Owned Enterprise (SOE) on behalf of the Ministry of Environment and Climate Change (above 100 m^3), the RBA (between 50 and 100 m^3) or the environmental department of the local administration (up to 50 m^3). For water use exceeding 100 m^3 per day (which will be the case of renewable hydrogen producers), EIAs need to be submitted as part of the license application file. Once industrial water users receive their water permit, they need to establish a water use certificate which includes a contract with the relevant RBA. This contract specifies the conditions for water utilisation, and the associated water service fee. External assessments have found that inconsistent procedures and standards have led to delays in issuing water permits (ADB, 2020[13]).

In addition, industrial users need to apply separately for a wastewater discharge permit. The permit is reviewed and issued either by the RBA (above 50 m^3 or if the wastewater contains hazardous substances as defined in the regulations) or by the environmental department of the local administration (up to 50 m^3). According to local stakeholders however, the process for obtaining a wastewater permit and the decision criteria used by the issuing authority are not clearly defined in the laws or in publicly information available. Reforms to the wastewater permitting regulatory framework will likely be necessary to allow for hydrogen production, due to the potential contamination or changing of water quality through desalination.

The revenues from water use charges are partly earmarked to finance environmental protection measures (German Development Institute, 2020[29]). Water abstraction fees differ among the 29 water basins, reflecting the ecological-economic value of the water resource in each basin, depending on the demand-availability balance and differences in economic activities of users (OECD and AWC, Forthcoming[33]). To increase revenues and further encourage wastewater treatment, the government adopted the Water Pollution Fee Law 2019. But the implementation of this law is facing challenges as local capacities and resources for measuring and monitoring pollution are limited and do not allow for systematic, reliable, and comprehensive sample testing on the ground. Nevertheless, the OECD National Water Policy Dialogue on Mongolia provides a series of recommendations for strengthening water use fees (OECD and AWC, Forthcoming[33]). In addition, water charges can be used to incentivise water-efficient behaviours, and this is something Mongolian authorities could explore moving forward, notably for energy and industrial activities such as renewable hydrogen production.

The Ministry of Environment and Climate Change is the sole entity able to issue a decision to carry out water exploration and research for industrial usage (as well as other usage), while project developers bear related costs. Local stakeholders interviewed for this report note that the possibility for projects developers to engage and participate in water exploration and research, in co-operation with Mongolian water authorities, would provide more flexibility in securing access to water for renewable hydrogen production.

To further develop the water allocation regime in the water-constrained South Gobi region, the OECD provides guidance on the key elements that a well-designed allocation regime should incorporate. These elements could assist in strengthening Mongolia's existing legal framework. A well-designed regime includes a clear legal status for all types of water resources, with competing claims clarified. It also encompasses a clear and enforceable abstraction limit accounting for in situ requirements and promotes sustainable use. Additionally, it involves legally defined volumetric water entitlements; and a robust water pricing system, typically under the form of abstraction charges, to encourage efficient and high-value water utilisation. The case of groundwater systems deserves particular attention in the design of allocation regimes due to their distinctive features (OECD, 2017[30]). The OECD also recommends to periodically perform health checks of water allocation regimes to ensure that they remain fit-for-purpose. The OECD National Water Dialogue in Mongolia provides a set of recommendations for improving the water management system and allocation regime of the country, including through financing instruments (Box 4.5). This timely exercise provides useful insights for ensuring access to water for future industrial uses, including within the developing renewable hydrogen sector.

Box 4.5. Strengthening water demand management and water allocation: highlights from the National Dialogue on Water in Mongolia

Water demand management policies can help to meet a growing need for water resources, reduce the risk of water depletion from unsustainable over-abstraction, and support making trade-offs between conflicting water uses. A range of tools contribute to water demand management, including economic instruments such as tariffs, water usage fees, and pollution charges, as well as water allocation systems comprising water information, permits, and water resources planning.

Even though the concept of '**water allocation**' is not formally embedded in policies, Mongolia has good practices of water allocation in place. The National Dialogue on Water highlighted several actions to further develop the water allocation system:

- **Provide a national framework** which embeds water allocation in the Integrated Water Management Plan, basin plans, and basin planning guidelines. The framework may cover principles, processes, methodologies, infrastructure needs, and capacity building.
- **Prioritise the introduction of water allocation systems in basins and aquifers facing temporary, structural or projected water stress**. This could also be a smaller hydrological unit, such as a sub-catchment. Having a good understanding of the water resource is a requisite.
- **Set and enforce long-term and short-term abstraction limits** through basin planning and permits that are aligned with the water balance and water availability. Water information will need to be strengthened, including a system to meter and monitor water uses.
- **Define a process for setting short-term abstraction limits during periods of water stress** (e.g., within a season). The process should consider the time needed for water users to adjust their water use and a sequence of priority uses during exceptional circumstances.
- **Develop grey or green water infrastructure that supports water allocation** through storage, treatment and water delivery, prioritising surface water use and wastewater treatment.
- **A water allocation regime based on the trade of water permits** ("water markets") may be viable in the long-term future when prerequisites such as a robust water planning, information and metering system are in place.

Water use and pollution fee are key elements of a water allocation system. Over the past years, Mongolia has made remarkable steps in strengthening the design and collection of water fees, which could be further enhanced:

- Water consumers, i.e. not-for-profit users such as households, government buildings and subsistence farmers, are exempted from paying the water use fee (Goverment of Mongolia, 2012[34]). To signal that water is a scarce resource and to contribute to the cost recovery of water management, **Mongolia could consider charging the water use fee to all users and consumers**. Special provisions, such as subsidies, should be in place for vulnerable groups such as low-income households and subsistence farmers.
- Water management can be improved through **better funding of RBOs**. This could be achieved by setting yearly transfer targets in the Integrated Water Management Plan.
- **Fee collection could be improved** by setting targets for fee collection in Integrated Water Management Plans, fostering cooperation and coordination between government entities on fee collection and inspection, and streamlining fee collection with the new digital permit system.

Source: Forthcoming report on the National Dialogue on Water in Mongolia (OECD and AWC, *forthcoming*)

The water regulatory framework is important to renewable hydrogen developers. If a hydrogen project is considered to be an industrial activity, then the developer will need to obtain a number of permits from the relevant authorities. These include Water Use Permit and a Water Use Agreement, as set out in the Mongolian Water Law, which are prerequisites for obtaining a Water Use Certificate. Broadly speaking, developers did not express any particular concerns about the regulatory framework for water use. However, some noted that the conditions for participating in the exploration of new water resources for potential project development were relatively restrictive. From the perspective of managing the hydrogen-water nexus, enhancing instruments aimed at fair water allocation among users based on the economic value of activities, and further internalising environmental costs by industrial operators, will play a key role. Moving towards a well-designed water allocation regime will help set a fair price for accessing water for renewable hydrogen production and incentivise water-efficient behaviours in the sector.

Mongolia's water infrastructure policy will play a key role in enabling – or not – the development of a domestic renewable hydrogen sector

As discussed earlier in this chapter, the problem of geographical mismatch between available surface water and centres of water demand in Mongolia extends beyond the issue of water for hydrogen production alone. It affects all other water users including agriculture, mining, and household consumption. On the other hand, as previously noted, if hydrogen developers are required to bear the costs of developing water infrastructure for renewable hydrogen production, the prospect of positioning Mongolia as a producing country would be jeopardised. However, if Mongolia's approach to developing water infrastructure is properly designed and implemented, it could help ensure access to water for renewable hydrogen production among other water uses.

Mongolia's Blue Horse umbrella initiative is the government's national flagship project for developing water and hydropower infrastructure in the country. This project aims to improve access to water across users including households, agriculture, and industry by bringing surface water resources located in the North to regions where water demand is high. A core objective is to transport water to the Gobi region via the construction of dams and reservoirs to supply water to the mining sector. Meanwhile, policymakers and the water scientific community in Mongolia are discussing the option of limiting or forbidding groundwater use for industrial activities starting in 2030. However, the Blue Horse initiative has received criticism from environmentalists, notably because of its plan to flood landscapes in Dauria, a transboundary steppe zone classified as world heritage by the UNESCO (Simonov, 2021[35]). Furthermore, some of the projects outlined in the framework of the Blue Horse project will involve cross-border dialogue with neighbouring countries.

Under the Blue Horse project, 33 technically feasible locations were surveyed on 12 major rivers in Mongolia. According to the results of the study, a total of 52 billion m^3 can be accumulated through the construction of dams and reservoirs in these 33 locations. Of the 33 locations, 10 were selected for their social, economic, and ecological importance. Projects in these 10 locations can generate up to 3.6 billion m^3 of water. One of the pilot projects in Mongolia stated in consultations with the OECD that they would not draw water from the Blue Horse project if it were built. This decision stems from concerns about potential social sensitivities associated with using inter-basin water transfers for industrial purposes as well as their confidence that their water needs could be sustainably met using local aquifers.

There are a couple of projects that are targeting future water supplies for industrial activities, including mining and renewable hydrogen production, in the South Gobi Desert. One such project is the Kherlen-Toono Project, which involves the construction of a 6 MW hydroelectric plant and a reservoir with capacity of a 1.6 billion m^3 near the Toono mountain in the Khentii province. This project is expected to result in additional water resources ranging from 120 to 130 million m^3, which will be transported to the Sainshand Industrial Park via a system of cast iron pipes and lifting pump stations. Another important project plans to collect underground water from Boor Khouvor, Dolood, and Tsagaansav to factories and mines in the Gobi region. This water will be stored in two 200,000 m^3 reservoirs located near Sainshand.

The Orkhon-Ongi project involves building a reservoir with a capacity of 598 million m^3 and a 2 to 4 MW hydroelectric power plant in the Uvurkhangai Province. It entails the construction of a port at the crossing of the Maikhan head of the Orkhon river to create a reservoir for regulating the river flow and transferring water to the Ongi river. Water from the Ongi river will be transferred to the Ulaannuur River through underground pipelines and then supplied to nearby towns and mining sites in the Gobi Province. The water distribution line for the project is approximately 535 km long.

The Blue Horse project has received endorsement from the Minister for Environment and is included in Mongolia's New Recovery Policy adopted by the Mongolian parliament in 2021. however, as of now, there is no clear plan for the projects' implementation and much uncertainty around the financing of this ambitious initiative. Due to its significant scale and likely environmental footprint, numerous discussions and studies are to be expected if the initiative moves forward. During consultations with the OECD, some local stakeholders from the industry, including renewable hydrogen project developers, expressed their scepticism or disinterest, indicating they would not utilise water from the project even if it were operational. The rationale for not drawing on water supply provided through the Blue Horse project is largely two-fold. On the one hand, project developers remain confident that local water resources will be sufficient for hydrogen production needs. On the other hand, developers are aware of the potential social and political sensitivities that might arise if they use an inter-basin water transfer to supply industrial projects, a position that was also echoed by established industrial enterprises in the country.

That the development of such a large-scale grey infrastructure project raises many challenges is not surprising. They are very large, complex, and costly projects, and in this case deal with a very socially sensitive resource. The Blue Horse project relies on inter-basin water transfers which typically bear a number of economic, environmental, and social challenges. Inter-basin transfers require the establishment of compensation mechanisms, which in turn require support from an adequate legal framework. These transfers can lead to large water losses through evaporation in desert areas. Additionally, in the case of Mongolia, they can lead to conflicts with border countries and generate disputes between sending and receiving regions. Korea faced a similar challenge when implementing inter-regional water transfer plans between the city of Busan and its neighbouring provinces. Busan aimed to address increasing water scarcity problems, while its neighbouring provinces opposed the plans due to concerns about their own water needs (OECD, 2018[36]). Such examples illustrate the importance of carefully preparing these plans, anticipating risks and consequences, and conducting a social dialogue to ensure buy-in from stakeholders including local communities. Furthermore, Korea's experience suggests that economic compensation for water transfer may become ineffective in the long term as the economic value of fresh water can increase over time due to the shift of balance between water availability and water demand (OECD, 2018[36]).

Alternatives, such as infrastructure for wastewater reuse, as discussed in section in 4.3.2, local water storage infrastructure, as already envisaged in Mongolia's plans, and nature-based solutions (Box 4.4) could provide economically and socially viable solutions addressing water scarcity problems in Mongolia's water-stressed areas. Through the National Water Policy Dialogue in Mongolia, the OECD suggests a number of avenues to develop and finance sustainable water infrastructure in the country (OECD and AWC, Forthcoming[33]). The development of water infrastructure is an essential condition for unlocking the potential of local renewable hydrogen production in Mongolia. As the national strategy for renewable hydrogen production moves forward, it will be necessary to consider the role of water infrastructure in facilitating access for this nascent industry. This aspect should be incorporated into the planning process to ensure effective implementation.

Box 4.6. Considering alternatives to large grey infrastructure

In addition to planning investments in large grey infrastructure for water storage, Mongolia could consider its current and future water storage capacities based on a mix of natural, built and hybrid water storage solutions, considering carefully their respective properties, advantages and disadvantages. Natural freshwater storage and nature-based solutions can form part of the water infrastructure solution. Wetlands, aquifers, lakes, soil, and other natural storage are also assets, that can contribute to meeting storage needs when adequately protected and maintained (Burke et al., 2023[37]).

Already, Mongolia is exploring managed aquifer recharge which can be a hybrid solution depending on its design features. Managed aquifer recharge is the intentional recharge of water to aquifers for subsequent recovery or environmental benefit. It can have significant potential to address the country's heavy reliance on groundwater and whilst also promoting circularity by treating and reusing wastewater (Fan et al., 2023[38]). A pilot project was supported by the Asian Development Bank, in Baganuur in Kherlen River Basin, and demonstrated the potential to enhance groundwater resources in Mongolia.

While natural storage solutions alone are unlikely to ensure regular base flows for household water supply, mining or industrial use, they can significantly contribute. Natural storage properties can be enhanced through increasing soil infiltration, via strategically located sand dams. Ensuring that natural storage solutions can deliver requires active and strategic management, based on careful analysis of soil structure and infiltration properties, to assess the ability to contribute to improved soil moisture and groundwater recharge. This may entail protecting land (particularly in headwaters) or engaging with farmers (including through systems of payment for ecosystem services) to incentivise them to adopt good management practices, notably in terms of grazing practices.

A range of other nature-based solutions can also be considered in Mongolia, such as the protection of watersheds. For example, the total economic benefit of drinking water from protected areas in Ulaanbaatar area has been monetised at 224.4 billion MNT (USD 165.1 million) by 2032 if investments in natural capital are made (Flores, Gonchigsumlaa and Surenjav, 2015[39]). Depending on costs of setting protected zones, this could make a compelling case to redirect revenues from Mongolia's water use fees towards nature conservation (2030 Water Resources Group, 2019[40]).

Source: Forthcoming report on the National Dialogue on Water in Mongolia (OECD and AWC, *forthcoming*)

References

2030 Water Resource Group (2014), *Mongolia: Targeted Anaysis on Water Resources Management Issues*, https://www.2030wrg.org/wp-content/uploads/2014/07/2030WRG_MONGOLIA.pdf. [32]

2030 Water Resources Group (2021), *Digital Water Platform: Development of groundwater monitoring portal using disruptive technology*, https://2030wrg.org/wp-content/uploads/2019/06/2030-WRG-Mongolia-Groundwater-Predictions-Report.pdf. [31]

2030 Water Resources Group (2019), *Building trust, growing resilience*, https://documents1.worldbank.org/curated/en/099113024053114453/pdf/P181147122840f0f21a2ec176210878c696.pdf. [40]

ADB (2020), *Overview of Mongolia's Water Resources System and Management*, https://doi.org/10.22617/TCS200202-2. [13]

Anna Nilsson, G. (2021), *Green Hydrogen Applications in Mongolia, Technology Potential and Policy Options*. [14]

Asian Development Bank (2020), *Making Water in Mongolia Available at the Right Time, at the Right Place, and in the Right Quality*, http://www.adb.org/publications/series/adb-briefs. [22]

Asian Development Bank (2020), *Overview of Mongolia's Water Resources System and Management: A Country Water Security Assessment*, Asian Development Bank, https://doi.org/10.22617/tcs200202-2. [3]

Becker, H. et al. (2023), "Impact of impurities on water electrolysis: a review", *Sustainable Energy & Fuels*, Vol. 7/7, pp. 1565-1603, https://doi.org/10.1039/d2se01517j. [7]

Beswick, R., A. Oliveira and Y. Yan (2021), "Does the Green Hydrogen Economy Have a Water Problem?", *ACS Energy Letters*, Vol. 6/9, pp. 3167-3169, https://doi.org/10.1021/acsenergylett.1c01375. [2]

Burke et al. (2023), *Lost Water: Challenges And Opportunities*, S&P Global, https://www.spglobal.com/_assets/documents/ratings/_research/101585883.pdf. [37]

Duranton, G. and D. Puga (2004), "Chapter 48 Micro-foundations of urban agglomeration economies", in *Handbook of Regional and Urban Economics, Cities and Geography*, Elsevier, https://doi.org/10.1016/s1574-0080(04)80005-1. [25]

Elsevier (ed.) (2020), *Dewatering mechanism of gasification fine slag by coupled mechanical force fields and its potential guidance for efficient dewatering process*, Fuel Processing Technology, No. 205, https://doi.org/10.1016/j.fuproc.2020.106459. [10]

Fan et al. (2023), "Groundwater Sensitivity to Climate Variations Across Australia", *Advancing Earth and Space Sciences*, https://doi.org/10.1029/2023WR035036. [38]

Flores, M., G. Gonchigsumlaa and K. Surenjav (2015), *Valuation of contribution of ecosystem services of the Gorkhi Terelj National Park to sectoral economic development*, https://www.researchgate.net/publication/291692737_Valuation_of_contribution_of_ecosystem_services_of_the_Gorkhi_Terelj_National_Park_to_sectoral_economic_development. [39]

German Development Institute (2020), *Forums, Fees and Data Flows: Coordinating Mining and Water Policy in Mongolia*, https://doi.org/10.23661/bp20.2020. [29]

Goverment of Mongolia (2012), *Law of Mongolia on Water*, https://faolex.fao.org/docs/pdf/mon154470.pdf. [34]

Government of Mongolia (2023), *Decision N°58 of the 8 February 2023*, https://legalinfo.mn/mn/detail?lawId=16758850092381. [23]

IEA (2023), *Global Hydrogen Review 2023*, https://www.iea.org/reports/global-hydrogen-review-2023. [1]

IEA (2023), *Tracking Clean Energy Progress*. [17]

IRENA and Bluerisk (2023), *Water for hydrogen production*, https://mc-cd8320d4-36a1-40ac-83cc-3389-cdn-endpoint.azureedge.net/-/media/Files/IRENA/Agency/Publication/2023/Dec/IRENA_Bluerisk_Water_for_hydrogen_production_2023.pdf?rev=7e9c17aa097340f4834f7df1c710ca72. [6]

Maddaloni, M. et al. (2023), "Exploring the Viability of Utilizing Treated Wastewater as a Sustainable Water Resource for Green Hydrogen Generation Using Solid Oxide Electrolysis Cells (SOECs)", *Water*, Vol. 15/14, p. 2569, https://doi.org/10.3390/w15142569. [20]

Madsen, H. (2022), *Water treatment for green hydrogen: what you need to know*. [41]

Marin, D. et al. (2023), "Hydrogen production with seawater-resilient bipolar membrane electrolyzers", *Joule*, Vol. 7/4, pp. 765-781, https://doi.org/10.1016/j.joule.2023.03.005. [19]

Muñoz, I. and A. Fernández-Alba (2008), "Reducing the environmental impacts of reverse osmosis desalination by using brackish groundwater resources", *Water Research*, Vol. 42/3, pp. 801-811, https://doi.org/10.1016/j.watres.2007.08.021. [42]

Nilsson, A. et al. (2021), *Green Hydrogen Applications in Mongolia: Technology potential and policy options*, New Climate Institute, https://newclimate.org/sites/default/files/2021/10/NewClimate_Green_Hydrogen_Applications_in_Mongolia.pdf. [4]

OECD (2024), *Environment at a Glance Indicators*, OECD Publishing, Paris, https://doi.org/10.1787/ac4b8b89-en. [12]

OECD (2024), *Mainstreaming Biodiversity into Renewable Power Infrastructure*, OECD Publishing, Paris, https://doi.org/10.1787/357ac474-en. [16]

OECD (2023), *Mining Regions and Cities in the Region of Antofagasta, Chile*, https://doi.org/10.1787/499ed299-en. [26]

OECD (2021), *Toolkit for Water Policies and Governance: Converging Twoards the OECD Council Recommendation on Water*, https://doi.org/10.1787/ed1a7936-en. [27]

OECD (2018), *Managing the Water-Energy-Land-Food Nexus in Korea: Policies and Governance Options*, OECD Studies on Water, OECD Publishing, Paris, https://doi.org/10.1787/9789264306523-en. [36]

OECD (2017), *Groundwater Allocation: Managing Growing Pressures on Quantity and Quality*, OECD Studies on Water, OECD Publishing, Paris, https://doi.org/10.1787/9789264281554-en. [30]

OECD (2016), *OECD Council Recommendation on Water*, https://www.oecd.org/environment/resources/council-recommendation-on-water.pdf. [28]

OECD (2015), *Water Resources Allocation: Sharing Risks and Opportunities*, OECD Studies on Water, OECD Publishing, Paris, https://doi.org/10.1787/9789264229631-en. [24]

OECD and AWC (Forthcoming), *Report on the National Dialogue on Water in Mongolia*. [33]

Saygin, D. and M. Lee (2023), *Financing cost impacts on cost competitiveness of green hydrogen in emergin and developing economies*, OECD Environment Working Papers, No. No. 227, OECD, Paris, https://www.oecd-ilibrary.org/fr/environment/financing-cost-impacts-on-cost-competitiveness-of-green-hydrogen-in-emerging-and-developing-economies_15b16fc3-en. [9]

Simonov, E. (2021), *A wetland refuge on the Russia-Mongolia border*, The Third Pole, https://www.thethirdpole.net/en/nature/landscapes-of-dauria-wetland-russia-mongolia-border/. [35]

Tonelli, D. et al. (2023), "Global land and water limits to electrolytic hydrogen production using wind and solar resources", *Nature Communications*, Vol. 14/1, https://doi.org/10.1038/s41467-023-41107-x. [5]

Tong, W. et al. (2020), "Electrolysis of low-grade and saline surface water", *Nature Energy*, Vol. 5/5, pp. 367-377, https://doi.org/10.1038/s41560-020-0550-8. [18]

U.S. Department of Energy (2013), *Reverse Osmosis Optimization*. [8]

UNICEF (2023), *Time is now to revolutionize water usage, Op-ed dedicated to World Water Week 2023*, https://www.unicef.org/mongolia/stories/time-now-revolutionize-water-usage. [15]

Water Services Association of Australia (2021), *Water Fuelling the Path to a Hydrogen Future*, https://www.wsaa.asn.au/sites/default/files/publication/download/Water%20fueling%20the%20path%20to%20a%20hydrogen%20future%20Nov%202021.pdf. [21]

WRI Aqueduct (2024), , http://aqueduct.wri.org. [11]

Notes

[1] Initial measurements from a bubbling well located at the southern tip of Bor Khuuvur indicate a mineralisation at 2,186 mg/dm3.

[2] Desalination requires only 0.072 kilowatt-hours of electrical energy (kWhel)/ kilogram of hydrogen (kgH2), while electrolysis consumes 53.8 kWhel/kgH2 (Tonelli et al., 2023[5]).

[3] Of note, IRENA does not consider upstream water consumption, i.e., water used for the production of fossil fuels consumed in hydrogen facilities.

[4] https://londonminingnetwork.org/2022/01/mongolian-herders-protest-at-rio-tintos-oyu-tolgoi-mine

[5] https://ikon.mn/opinion/32fj

5 Recommendations

This chapter presents the Recommendations of the OECD report 'Towards a renewable hydrogen strategy in Mongolia'. The extractive-fuelled growth of Mongolia poses challenges for the long-term sustainability ambitions of the government and raises the importance of policy support for technology and investment decisions that can align industrial growth with low carbon outcomes. Renewable hydrogen has emerged as one potential technology option for the long-term sustainability of Mongolia's industrial sector. The domestic production of renewable hydrogen could also contribute to a range of other government ambitions, such as export diversification and industrial development. Yet, under current conditions, attracting the investment necessary to develop a renewable hydrogen industry is challenging. The recommendations of this report explore how the government can support the emergence of a renewable hydrogen industry in Mongolia, and which can contribute to the broader long-term sustainability and socio-economic development objectives of the authorities.

Context for the recommendations

The recommendations of this report are designed to support policymakers in Mongolia with the development of a renewable hydrogen strategy. The aim of the recommendations is to align decision-making on hydrogen-related development and its supportive industry with the sustainable, low carbon growth of the Mongolian economy. The recommendations draw on the analysis presented in this report, OECD experience in the areas covered by the report, and the extensive consultations that the OECD Sustainable Infrastructure Programme in Asia (SIPA) held with public and private stakeholders in Mongolia in 2023-24. The recommendations and analysis that underpin them were discussed and validated with the Ministry of Economy and Development of Mongolia at a workshop in Ulaanbaatar in April 2024 which also brought together representatives of the Ministry of Energy and of the private sector including the Mongolian Hydrogen Council.

Hydrogen pilot projects in Mongolia for renewable hydrogen development demonstrate that there is private sector interest. Any strategy should therefore look to empower the private sector to responsibly and sustainably explore opportunities for renewable hydrogen development, to leverage experience from early-stage projects to further the government's sustainable economic development agenda, while not closing the financial, policy and innovation space to alternative technologies that could support a low carbon transition in Mongolia. This opportunity to engage in a public-private industrial policy dialogue on renewable hydrogen is important to Mongolia's future sustainable economic growth, regardless of whether it leads to great commercial success. Even in the absence of such success - which is never guaranteed when it comes to innovation - it can and should be seen as a chance at getting the (policy) *process* right, as Dani Rodrik puts it (Rodrik, 2004[1]), and build institutional knowledge and capabilities that will serve for the needed transformation of Mongolia's growth model. It will also enable stakeholders to learn about the potential use of renewable hydrogen technology whether Mongolia produces it or imports it in the future.

The nature of any low carbon transformation is systemic, and as such the recommendations generally draw on the entire report rather than specific thematic chapters (Figure 5.1). While there are exceptions to this, notably in the recommendations on water, it means that the findings of several chapters of the report may be present in a given set of recommendations, even when these relate to targeted and well-defined policy actions. Other recommendations, particularly those concerning strategic planning and steering, are deliberately transversal. Taken together, the recommendations may act as a roadmap for the government as it moves forward with the development of a renewable hydrogen strategy.

The recommendations are grouped into three thematic categories. The first set of recommendations (Recommendation 1 and 2) concern the process of developing and implementing a strategy for renewable hydrogen development in Mongolia. The second set of recommendations (Recommendations 3-5) look at how to improve the framework conditions for industrial decarbonisation in Mongolia, including creating demand for low carbon technologies such as hydrogen, and improving the competitiveness of Mongolia as a destination for foreign direct investment in low carbon industrial activities. The third set of recommendations (6-8) look at how to ensure the development of infrastructure – both hard and soft – that can ensure a sustainable use of the country's natural resources in the pursuit both of a renewable hydrogen strategy and the country's sustainable economic development more broadly.

A key consideration for the government of Mongolia in designing and implementing a national strategy for the development of renewable hydrogen will be sequencing. Many of the national strategies that have informed the analysis of this report reflect the ambitions of governments to establish their countries as first movers in the production and use of renewable hydrogen. While this may be the approach that the government of Mongolia chooses to take, the recommendations have nevertheless been structured in such a way as to be adaptable to different timelines and ambitions. In practice, this means that certain recommendations will be of greater relevance to the Mongolian authorities in the short term should they decide to establish the country as a first mover in the production of renewable hydrogen. At the same time, a larger share of the recommendations will be relevant to policymakers even if they adopt a less directional and more cautious approach to hydrogen development; indeed, they will be of relevance to policymakers even if they do not develop a renewable hydrogen strategy at all. This is because while all of the recommendations included in the report will be necessary for a renewable hydrogen industry to develop in Mongolia, their implementation will support the development of other low carbon technologies, capacities, and infrastructure necessary for the decarbonisation of Mongolia's economic growth.

Figure 5.1. Towards a renewable hydrogen strategy for Mongolia: Overview of the report (reminder)

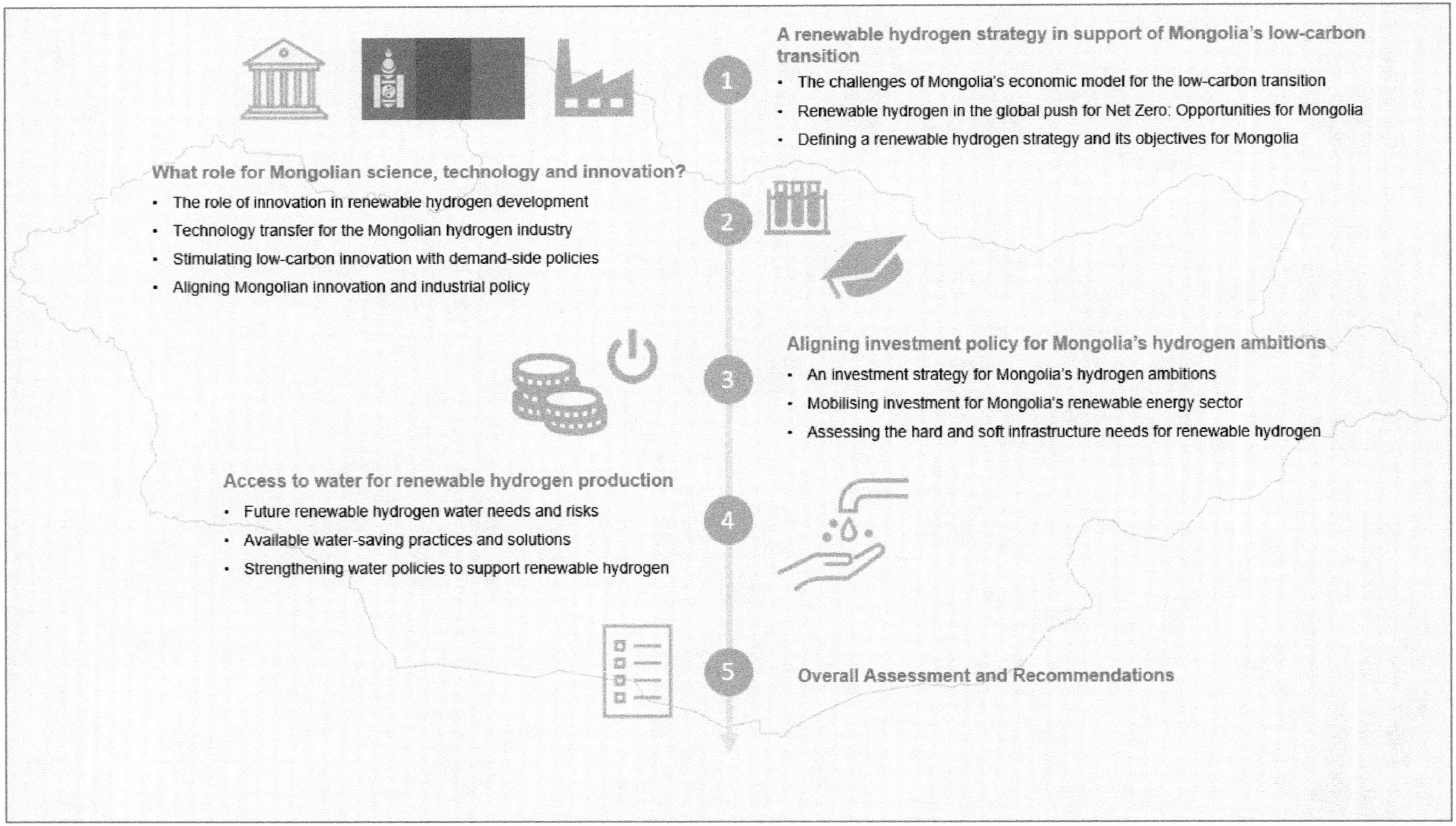

Figure 5.2. Towards a renewable hydrogen strategy for Mongolia: Overview of the recommendations

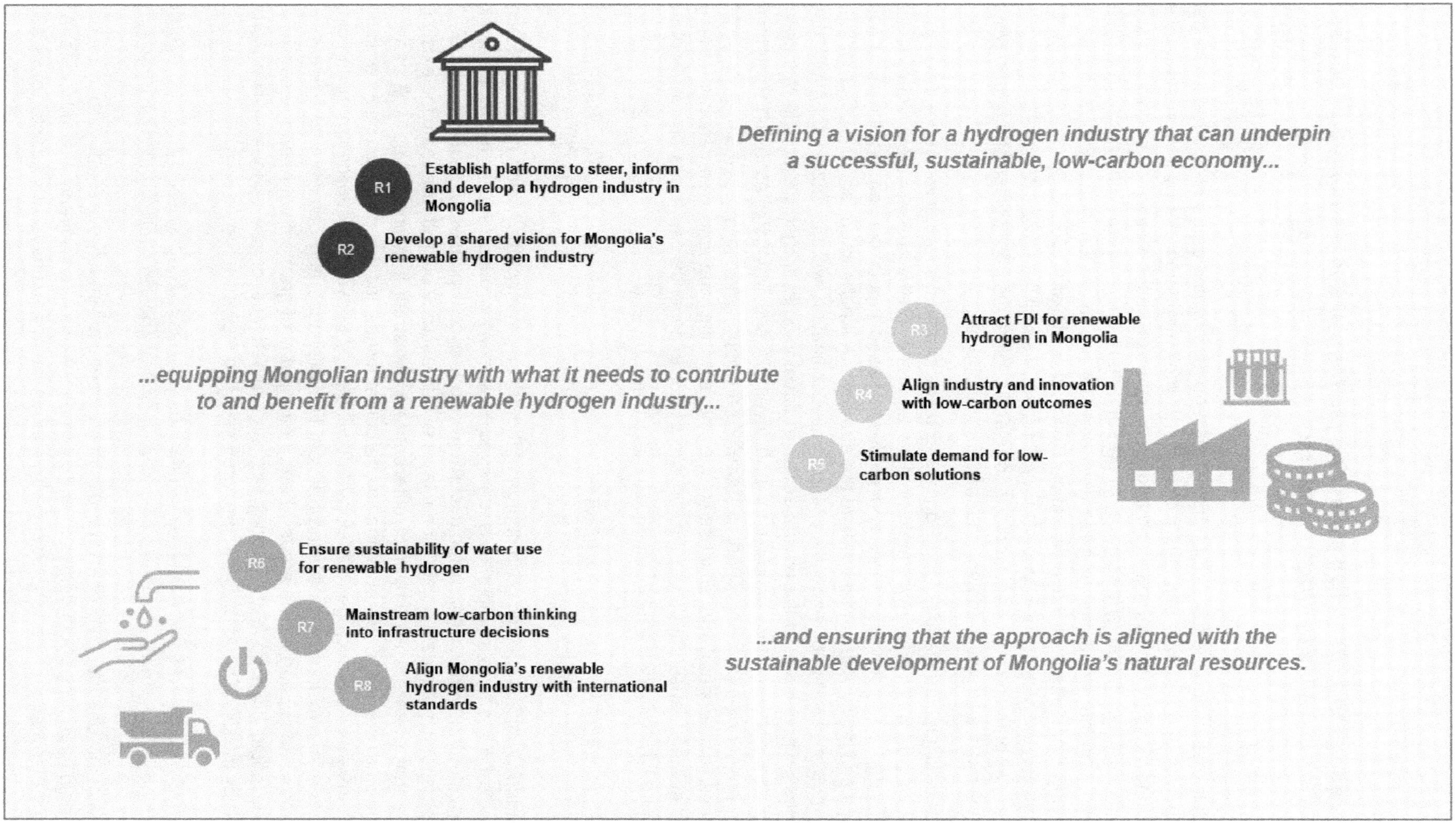

Recommendations

Recommendation 1: Establish a steering platform and implementation bodies for Mongolia's renewable hydrogen strategy

Overview

Successfully navigating the low carbon transition requires aligning policy areas that in the past have often been disparate and unconnected; this is true of the implementation of a hydrogen strategy as it is of any low carbon technology development and adoption that involves the interaction of domains such as energy, industry, finance, education, and science. At present, there are conversations ongoing in different areas of government and industry concerning the development of a hydrogen industry in Mongolia. For example, the Standing Committee on Industrialisation Policy, the Ministry of Energy, the Ministry of Mining and Heavy Industry, the Ministry of Economic Development, as well the private sector-led Mongolian Hydrogen Council, are all pursuing important work to scope and explore the possibilities of a future hydrogen industry in Mongolia but are largely doing so independently of one another.

The successful implementation of the Concept and any eventual strategy will depend on the ability to align and steer various public and private stakeholders, as well as on the ability of those charged with its steering being provided with relevant, timely and evidence-based inputs concerning the evolution of different technology and policy issues that relate to the strategy. Given the importance of different line ministries in directing public investment and policy support, coordination will be critical to avoid duplication and wastefulness. The interest of the private sector in developing hydrogen projects in Mongolia is a positive indication that entrepreneurs, innovators, and investors see Mongolia as a high-potential location for hydrogen project development. There is therefore an opportunity for policymakers to integrate projects' experiences, challenges, and successes into the steering and implementation of the government's hydrogen strategy, ensuring bottom-up, private sector experience can focus government intervention to areas where it is needed most.

R1.1: Establish a cross-government, public-private Steering Committee for the low carbon transition in Mongolia

The Steering Committee would be in charge of the development of the Vision set out at Recommendation 2 and could be fed by a number of issue-specific working groups (e.g., a hydrogen working group, a technology-neutral working group for decarbonisation, a coal phase-out working group). The Steering Committee should not create an additional layer of bureaucracy but rather streamline existing platforms. It is notable that there are a number of different line ministries in Mongolia as well as cross-ministerial committees that are separately pursuing low carbon activities, including on hydrogen; ensuring an aligned approach to renewable hydrogen development will be key to avoiding duplication, giving clarity to project developers and investors, and allowing relevant line ministries and other public agencies to focus their interventions and support on the areas where their input is of greatest value and necessity. The Steering Committee could build upon the participants and experience of the various working group meetings organised by the OECD and UNDP in preparation of this report.

R1.2: Conduct public and private sector awareness raising exercise on government ambitions for decarbonisation and the role of low carbon fuels such as renewable hydrogen

There is a highly unequal level of awareness of the economic imperative for decarbonisation in Mongolia. While certain large firms, both domestic and international, have an advanced and long-term vision on the importance of investments to support decarbonisation, other firms have significantly lower awareness. The ability of a hydrogen industry to contribute to a just transition will depend on creating opportunities for firms – in particular SMEs – in existing and future industrial value chains linked to the use of low carbon technologies and fuels such as hydrogen. A similar dynamic is clear in the public sector, including in line ministries whose input will be necessary for any strategy to be successful and whose broader responsibilities will be significantly affected by the low carbon transition. There is therefore a pressing need to raise awareness at a technical and policy level across the public and private sector on the rationale of public policy support for low carbon technology, as well as the commercial opportunities and challenges relating to the government's ambitions for low carbon economic development.

R1.3: Raise awareness of the opportunities and challenges of hydrogen development at the level of regional government

There is a specific need to raise the awareness of renewable hydrogen projects and government policy objectives in this area at the level of regional administrations. Renewable hydrogen development projects have the potential to significantly contribute to regional development priorities, including infrastructure development and job creation; this is particularly important in regions that may over the longer term be affected through job loss or displacement owing to a future coal phase-out, or at least a coal economy decline. At the same time, the use of water resources and land – for example in the development of renewable energy infrastructure for electrolysis – may have significant localised impacts on communities. There is a clear rationale for increasing the involvement of regional governments in the consultation, design and implementation of renewable hydrogen projects to ensure that such projects contribute to local development needs and are socially and politically viable at the local level. This process also runs in the other direction, since many of Mongolia's regional governments have deep expertise and experience in managing large industrial projects; in this sense, there is much that can be derived from the experience of regional government to inform national decision-making on low carbon industry development, including for renewable hydrogen.

R1.4: Establish a public-private working group to map infrastructure needs for low carbon industry development

There is a significant level of uncertainty surrounding the infrastructure needs for hydrogen exports from Mongolia, with much dependent on the location of production and domestic users, the proximity to the Chinese border, clustering of production and related energy infrastructure, and infrastructure developments in China. Similarly, there is uncertainty surrounding the low carbon industry infrastructure needs more generally, both in terms of new infrastructure, retrofitting of existing infrastructure, planning, permitting and location of industry development and its associated infrastructure, etc. While different working groups are envisaged under R1.1, the cross-sectoral implications for infrastructure development and the associated financing needs justify a particular focus on these questions. The lessons derived from the working group could significantly improve the quality and availability of information available to both the private and public sector in the planning of industry investment projects.

Recommendation 2: Define a shared vision for a renewable hydrogen industry in Mongolia

Overview

The Government of Mongolia is considering the development of a renewable hydrogen industry and is exploring how best to support the development of such an industry in line with its broader socio-economic, industrial and sustainability objectives. Drawing on the experience of other economies that are exploring or in the early stages of implementing national strategies for renewable hydrogen development, Mongolia should approach the development of a national strategy in a clearly sequenced manner. The first stage of this approach should be establishing a socially and politically acceptable vision for what a renewable hydrogen industry could and should provide for Mongolia and its citizens, and how such an industry could contribute to the sustainable development of the country, including a just transition away from fossil fuel production and use. This stage should include a clear and evidence-based analysis of key early-stage risk factors associated with renewable hydrogen development – off-taker risk and finance risk – that can affect the viability and affordability of hydrogen projects, particularly in emerging markets. The findings of these exercises could then inform the development of a national strategy for renewable hydrogen development.

R2.1: Develop a shared 'Vision' for a renewable hydrogen industry in Mongolia

The Vision would act as an initial stage towards the development of a hydrogen strategy in Mongolia and would set out a vision of what a renewable hydrogen industry in Mongolia could contribute to the country's socio-economic, environmental and climate policy objectives. The Vision could also outline the key responsibilities and needs of the public and private sector in the design and implementation of a renewable hydrogen industry. The Vision would clearly detail the risks and opportunities involved in directing public support towards the establishment of such an industry. Concretely, the Vision could:

- Set clear targets for how hydrogen should contribute to decarbonisation in Mongolia, together with targets from other renewable technologies.
- Mapping of resources and data collection to identify best locations for renewable hydrogen production.
- Set out how the development of a renewable hydrogen industry could contribute to social and regional inclusion, for example through job creation, industrial investment, or upgrades to energy and power infrastructure.
- Clearly establish the environmental and resource implications of a hydrogen industry in Mongolia at different magnitudes of scale.
- Identify the potential domestic demand for renewable hydrogen including for supporting its early-stage development. In particular, this should be linked with the identification of industrial products where additional value and differentiation could be created by embedding low carbon fuels. Experience from other potential export-oriented hydrogen producers indicates that a number of countries are looking to embed renewable hydrogen directly into other industrial value chains, rather than exporting the hydrogen itself. In Mongolia, given the planned expansion of copper mining, there may be scope to couple the development of a renewable hydrogen production industry with the development of a new, forward-thinking, strategically important and highly differentiated product such as low- or zero-carbon copper.
- Align and be coherent with government objectives for decarbonisation – which are being developed in a Long-Term Vision exercise supported by international partners - and its commitments under Mongolia's Nationally Determined Contribution.
- Establish a clear role for the local science, technology, and innovation system in the development of a renewable hydrogen industry.
- Set clear short-term technology-neutral targets and long-term technology-specific targets, with a roadmap for how these will be achieved.

Recommendation 3: Improve Mongolia's attractiveness as a foreign direct investment destination for renewable hydrogen development

Overview

The success of Mongolia's renewable hydrogen ambitions will in large part depend on the ability of the country to attract investment from external sources, especially in a context where other emerging economies are also competing to attract FDI into their own nascent hydrogen sectors. A key role for the government should therefore be promoting Mongolia as an attractive destination for low carbon investment, using both financial and non-financial policy interventions to de-risk such investments where appropriate, addressing issues in the broader business environment that may undermine investor confidence, and utilising the existing policy and regulatory framework for investment to create clear economic signals that the government is committed to a low carbon industrial future.

R3.1: Integrate a low carbon investment mission into the Mongolian Investment and Trade Agency

The government could consider integrating a low carbon investment attraction mission into the mandate of its pre-existing investment promotion agency (the 'Investment and Trade Agency of Mongolia' under the Ministry of Economy and Development) and use trade representatives in diplomatic delegations to increase the awareness of Mongolia's renewable hydrogen development ambitions. Building on the projects identified in this report, the Agency could also consider defining a suggestive pipeline of projects for foreign investors and facilitating dialogues between the projects and potential investors.

R3.2: Identify and address business climate issues that may negatively affect investor confidence in Mongolia as a destination for renewable hydrogen development

There are a number of issues in the broader business climate that may negatively affect investor confidence in Mongolia, some of which have been discussed in this report, such as the ongoing perceived dispute surrounding power purchasing agreements in the renewable energy sector. Given the high risk involved with renewable hydrogen projects, such unresolved issues may lower investor confidence in the reliability and stability of Mongolia's otherwise open and liberal legal and regulatory framework for investment. Ensuring that the government is able to solve issues and disputes swiftly and transparently in the business climate will be critical to increasing investor confidence not only in the nascent hydrogen industry but the broader energy and infrastructure necessary for Mongolia's low carbon transition. Fundamentally, the importance of attracting foreign investment for low carbon investment projects should redouble the need to improve the broader investment climate.

R3.3: Consider import duty and tax relief for enabling technologies for low carbon and sustainable industries

There are a number of horizontal legal provisions already in force in Mongolia that might support investments to support a low carbon industrial transformation and low carbon innovation more broadly, notably Art. 12.1.4 of the 2013 Law on Investment (conditions for non-taxable investments for the production of export-oriented innovative products to finance innovation) and the 2012 Law on Exemption from Customs and VAT (customs and VAT relief for the import of equipment for innovative products). Mongolia could consider applying a clearer low carbon taxonomy to such provisions, for example by extending them to capital goods necessary for hydrogen production, or to the inputs necessary to build transmission and distribution infrastructure. This could send a clear signal to the market that the government is privileging investment in areas that support its low carbon objectives, and encourage incumbent investors to modify their own investment activities and plans.

R3.4: Conduct public-private dialogues on specific regulatory and policy barriers to renewable hydrogen production and applications in Mongolia

By virtue of being a frontier technology, there is a significant degree of uncertainty regarding the specific regulatory and policy barriers to the development – and use – of renewable hydrogen beyond its traditional industrial applications. As discussed in Chapter 3, many of these uncertainties relate to certification and standardisation requirements for hydrogen production and its derivatives, as well as industrial safety standards for its use and transport. As private projects continue to develop and as the government advances with the articulation of its own hydrogen ambitions, a public-private working group with a specific mandate to identify and address regulatory and policy barriers to hydrogen production and its application in Mongolia could further strengthen investor confidence and demonstrate the government's willingness to facilitate the emergence of a high-potential green industry.

R3.5: Accelerate reforms to improve attractiveness of renewable energy sector, balancing the need to create incentives for the private sector with long-term sustainability of public finances

The renewable energy sector in Mongolia must be developed if the country is to meet its commitments set out in its NDC, regardless of whether Mongolia decides to develop a renewable hydrogen industry or not. Yet, at the same time, the country's vast renewable potential is one of its main comparative competitive advantages as a potential hydrogen producer, and indeed the potential to produce low cost, clean energy could support the decarbonisation of the Mongolian economy and even attract energy-intensive industrial investors looking to decarbonise their production. While Mongolia's renewable hydrogen do not change the need to develop the renewable energy sector, they do add additional economic rationales for doing so. In addition, while the current electricity grid is too carbon-intensive to provide power for renewable hydrogen production, should that change and if the rules of additionality can be met as discussed in Chapter 3, then it will only increase the competitiveness of hydrogen production in Mongolia. Understanding and addressing the barriers that continue to restrict – or dissuade – investment in the sector must be a priority for the government going forward. This includes accelerating the ongoing reforms to regulated electricity prices (increasing the returns of dedicated renewable projects while also increasing the rentability of hydrogen projects with proprietary renewable energy infrastructure that can sell surplus electricity to the grid), undertaking reforms to Mongolia's feed-in tariffs and ensuring that their attractiveness for investors is properly weighed with public financial sustainability, and addressing investor concerns in the power purchasing agreement framework.

Recommendation 4: Align Mongolia's industrial development with low carbon outcomes

Overview

The policy and regulatory framework to incentivise investment in Mongolia's low carbon transition remains at an early stage of development. Firms lack the knowledge, technology and incentives to innovate in and adopt new solutions that can align their industrial activities with low carbon outcomes. There are however a number of reforms and policies that could significantly raise the contribution of the private sector to low carbon innovation and technology adoption. Such reforms could be technology-neutral, giving directionality to firm-level investment and innovation decisions in a way that aligns with low carbon transition objectives without pre-determining which solutions they explore. There are two key areas where Mongolia can substantially increase the prospects for the development of a renewable hydrogen industry whilst ensuring that the development of such an industry benefits the local industrial ecosystem.

The first is to make it more attractive for firms and the innovation system that supports them to direct their investments and resources towards activities that support low carbon outcomes. This means sending signals through the innovation and industrial policy framework that there is a clear economic rationale to invest in low carbon technologies, innovation, and capacities. At the same time, there is a need to make it not just more attractive but also easier for firms to undertake those investments.

The second key area for action is to make it easier for industrial firms to access the technologies and innovations that can support their decarbonisation and access new opportunities in low carbon value chains. An inclusive and durable low carbon transition for Mongolia's industry will require equipping firms and workers with the skills, knowledge, and capacities to maximise the opportunities involved in low carbon technology adoption and innovation. Doing so will ensure that Mongolian firms, both those that currently operate and those which the government hopes to see emerge as its industrialisation policy advances, are able to competitively position themselves in new, emerging green industries, while also adapting their current activities to changing regulatory and consumer demands. On one hand, this requires supply-side innovation and technology policies to facilitate the transfer of technologies and know-how to firms and workers. On the other hand, it requires demand-side, market-shaping policies that create demand for low carbon innovation in the domestic economy.

R4.1: Establish a public low carbon innovation policy and technology laboratory for Mongolia

The laboratory would act as a focal point for low carbon projects in Mongolia, centralising information on their focus and aims, it could monitor the effectiveness and status of various technology-specific and technology-neutral barriers to low carbon innovation, investment and technology adoption. To avoid duplication, the laboratory could be embedded within an existing project development agency, such as the New Recovery Policy Accelerator. This would allow the early-stage support of a strategically important area of industrial and low carbon development to be embedded in a high-impact, dynamic institution with a cross-sectoral mandate.

R4.2: Consider the introduction of a low carbon R&D tax credit for the private sector and state-owned enterprises

The introduction of a low carbon R&D tax credit would send a clear signal to firms – both private and state-owned – that the government is ready to place a premium on investments that might support the low carbon transition of industry. While such credits are generally horizontal and untargeted, Mongolia could consider applying a level of conditionality – such as that set out in Mongolia's Sustainable Development Goals Financing Taxonomy – to direct firm-level innovation activities towards low carbon objectives.

R4.3: Mobilise development and domestic finance to support directed innovation financing tailored to Mongolia's low carbon needs

OECD research suggests that there is a greater justification in using targeted, grant financing for innovation in emerging technologies with a strong sustainability potential but which remain far from the market (OECD, 2023[2]). Proposals such as those made by the World Bank Energy Sector Management for Development (ESMAP), the OECD, the Global Infrastructure Facility and the Hydrogen Council to establish 'lighthouse' projects for hydrogen production in emerging and developing economies, which can help to crowd in and scale development finance for hydrogen development, are indicative of the financing opportunities that could be available to Mongolia should it advance with its hydrogen ambitions (OECD, World Bank, ESMAP, GIF, Hydrogen Council, 2024[3]). Mongolia already has experience with securing development finance for sustainability objectives – such as through its use of Japan's Joint Crediting Mechanism – and could consider a more explicit policy of orientating development finance towards innovation in low carbon areas that may have a strong social impact but limited commercial viability. There are also a number of concessional green finance instruments available domestically in Mongolia (see, for example, the activities of the Mongolian Green Finance Corporation), but these mostly relate to the residential sector. There is scope to expand the availability of concessional finance instruments with an explicit goal of low carbon technology adoption in industry.

R4.4: Establish a technology transfer unit within Mongolia's higher education and research institutions with a mandate to promote and enable low carbon technology transfer

A dedicated unit, either with a higher education institution (HEI) affiliation or operating across a consortium of institutions, to identify opportunities for domestic and international low carbon technology transfer for domestic industry could overcome informational asymmetries between Mongolian industry and science and research, and facilitate opportunities for domestic industry to capitalise on opportunities in emerging green industries or in the low carbon transitions of their current area of industrial activity. The advancement of the unit's activities could be made part of the explicit mandate of local HEIs and research institutions.

R4.6: Explore options for academic-industry placements in emissions-intensive sectors to facilitate technology transfer

In many firms, particularly SMEs and certain SOEs, there is often a lack of internal capacities – financial, knowledge and skills – to adopt and use new technologies that might support firms with productivity, innovation and their decarbonisation. The government could consider a specialist academic placement programme, whereby doctoral students or suitably experienced graduate students are encouraged or academically rewarded for placement periods within a firm with the aim of supporting their uptake and use of a specific technology or other form of innovation with a view to addressing a low carbon objective.

Recommendation 5: Define a role for the government in stimulating demand for low carbon technologies

Overview

One of the major barriers to developing any low carbon industrial project in Mongolia is a lack of demand for low carbon goods and services in the country, both in terms of the market size and market preparedness. While the domestic industrial market is small, it is growing, and the government has a clear commitment to developing many industrial activities in which the integration of low carbon fuels and technologies will be necessary for their sustainability and competitiveness. While it is also true that there is scope for external markets to become drivers of low carbon industrial output in Mongolia, experience from the country as well as from throughout the region points to the importance of domestic off-take in reaching final investment decision in low carbon industrial projects.

There is an emerging consensus across the OECD that the many supply side policies for low carbon technologies that have been implemented in recent years have not yet been met with a commensurate commitment of demand side policy interventions. Given that stimulating demand for low carbon goods and services depends in large part on addressing the market failures that have created the need for such goods and services in the first place – namely, the failure of industries and consumers to price in the negative environmental and climate externalities of a business as usual scenario – there is a clear economic and climate rationale for governments to play a role in accelerating the uptake of low carbon technologies, particularly in hard-to-abate sectors. The scope of that role will nevertheless be determined by the financial capacities of the government in question, which in emerging economies are generally lower than in most OECD members.

While there is therefore a need for the Mongolian government to play such a role in the country, there is also a need to define it in terms that make effective use of more limited resources. In this regard, there are two main areas that policymakers in Mongolia could consider for action. The first is to identify the extent to which the 'business as usual' policy and regulatory framework acts as an implicit dampener on domestic demand; how, for example, policy and price signals around coal usage in industrial and energy sector development translates into investor sentiment regarding low carbon markets in Mongolia as well as the competitiveness of those low carbon industrial activities relative to highly polluting and emissions-intensive alternatives. The second is identify what interventions the government can proactively take to stimulate demand for low carbon industrial technologies in the industries that are in greatest need of low carbon solutions for their future sustainability and competitiveness.

R5.1: Conduct a socio-economic assessment on the impact of fossil fuel subsidies on investment in and demand for low carbon technologies

While it is important to ensure that subsidy reform does not create a distortionary imbalance between energy supply and demand, which risks triggering price increases that would hit the poorest segments of society the hardest, most international organisations and research agree that a rationalisation and gradual reduction of fossil fuel subsidies are fundamental ingredients of a successful low carbon transition. In Mongolia, the heavy subsidisation of coal in particular sustains very low electricity prices which in turn may weigh on the investment attractiveness of the renewable energy sector and technologies that could support the low carbon transformation of industry. An inventory exercise of all direct and indirect fossil fuels would be an important starting point in this regard.

R5.2: Design a policy toolkit to stimulate demand for low carbon technologies in Mongolia's industrial sector

Mongolia lacks a policy toolkit to stimulate demand in low carbon technologies. While certain provisions do exist within the fiscal and customs framework to lower capital costs for renewable energy projects (R3.3), these are not strategically applied; beyond these provisions, there are no 'market shaping' policies that are relevant to low carbon technologies. There are a number of policies that the government could consider as it develops a strategy not just for the production of renewable hydrogen, but also its use – and the use of other low carbon technologies – within Mongolian industry. These can be technology specific mandates and quotas; the replacement of 'brown' hydrogen with renewable hydrogen in existing industrial activities, or targets to use a defined amount of renewable hydrogen in certain activities, such as public transport or mining. Should they be financially sustainable and aligned with the government's strategic aims, the government might also consider direct subsidies for the use of renewable hydrogen in hard-to-abate sectors through tax credits, subsidies for the capital and operational expenditure for renewable hydrogen production and use, or more technology neutral instruments such as carbon contracts for difference.

R5.3: Use low carbon conditionalities in public procurement, including in Mongolia's state-owned enterprises

Given the particularly large footprint that the state retains in some of the country's hard-to-abate sectors, public procurement is one area where the government could immediately stimulate demand for low carbon technologies. Mongolia has explored the use of 'green' procurement for a number of years, but the government has yet to leverage the ability of the significant purchasing power of the state and state-owned enterprises to create demand for low carbon technologies and send clear market signals to industry and investors about the government's low carbon objectives. A more explicit low carbon approach to public procurement could support the adoption of technologies at an early stage of commercialisation in strategic sectors, including mining and energy.

R5.3: Support the emergence of a local hydrogen production industry through bilateral trade agreements with key countries for off-take contracts

Given the importance of external off-takers for the economic viability of Mongolian renewable hydrogen production, there may be a role for the government to play in facilitating and supporting potential external off-take contracts through bilateral agreements. For example, in the case of China, where there is the potential for significant renewable hydrogen demand in the coming years, there may be a role to play for the government in establishing agreements with the relevant purchasing authorities or enterprises, and by extension giving greater clarity to hydrogen producers in Mongolia on what opportunities there might be for export to the Chinese market. This type of international dialogue may also be important for the development of the supportive export infrastructure that will have to be built should Mongolia export hydrogen in the future.

Recommendation 6: Ensure the sustainable use of Mongolia's water resources for hydrogen production

Overview

Water availability for renewable hydrogen production is a critical concern for policymakers and for potential investors in Mongolia, and more broadly in Central Asia, a region facing water scarcity issues bound to be aggravated by climate change. In Mongolia, the problem takes the form of a spatial mismatch between the water resource availability, concentrated in the North of the country, and centres of water demand, largely in the Ulaanbaatar area and the South Gobi desert. As the South Gobi region is seen as the most promising location for renewable hydrogen production means, there is a high risk that the development of this sector at scale will aggravate water stress in this region.

Available data, which relies on a few studies based on laboratory conditions, allow to estimate that producing 1 kg of hydrogen in the South Gobi desert of Mongolia would require to abstract just above 40 litres of groundwater, out of which 70% (about 28 litres) would be consumed while the rest (12 litres) would be discharged under the form of brine (from water pre-treatment required for the electrolysis) and warm water (from the cooling process post-electrolysis). Although the required amount of water for developing pilot projects remain relatively limited when compared to consumption in other sectors, the development of a renewable hydrogen sector at scale does can lead to large volumes of water consumed. Under these estimates, a 1 MT capacity industrial development of renewable hydrogen production (a relatively standard mid-term capacity target considering international approach to hydrogen production plans) would lead to a water consumption of about 30,000 thousand m3/year, which is close to the forecasted water needs of the mining industry in the South Gobi region by 2030, mining being the primary industrial user in this region.

Adopting a "nexus" approach to hydrogen and water management, i.e. an approach acknowledging the inextricable link between the two sectors, cognisant of the fact that policies in one area will affect the other, can help address risks resulting from this relationship and uncover opportunities for synergies. Uncertainty regarding sustainable access to water at the horizon of investments (which, in the case of renewable hydrogen projects, are long-term), can impede the attraction of financing into the sector. A clear and reliable water policy framework that manages water disruption, environmental risks and potential social conflicts around water will contribute to de-risking investment in the sector. On the other hand, envisaging water and hydrogen linkages can lead to economic opportunities, for example through the development of a circular economy approach in which wastewater is used for renewable hydrogen production.

R6.1: Integrate water sustainability as a central pillar of a national renewable hydrogen strategy

Mongolia needs to make water considerations central to its approach to hydrogen development. There is a need to pursue the analysis undertaken in this report to understand present and future water resource availability, as well as present end future water demand across uses, and assess what can be allocated to hydrogen production, based on its expected value for the national and local economies. This discussion, as well as the mainstreaming of water efficiency requirements into an approach that aims to develop the renewable sector, can be done through a dedicated working group under the steering platform that this report is recommending for developing a renewable hydrogen strategy. Actions for integrating water sustainability into renewable hydrogen development can entail:

- Including the effects of an industry scale-up on water sector in the discussion about renewable hydrogen production targets. In the short-term, the effort can focus on securing access to water for enabling the development of pilot projects while establishing the conditions for ensuring that it does not create tensions with other uses and does not contribute to unsustainable water resource depletion. In the longer run, there is a need to carefully envisage the water volumes that can effectively and sustainably be engaged in renewable hydrogen production and understand how this can constraint the scale at which a renewable hydrogen sector can develop. Considering the linkages with national industrial and regional and local development plans and strategies can make sense, as plans for reducing specific water-intensive industrial activities, for example, can have an impact on the total water availability for other purposes.
- Monitoring closely the anticipated water needs of renewable hydrogen projects under development. Most available data rely on a limited number of cases reflecting laboratory rather than real-life, at scale conditions. Monitoring effective water parameters of real-life projects, as pilot projects develop in Mongolia, will help build knowledge reflecting local conditions.
- Developing a database of hydrogen-water metrics used for domestic standards, benchmarking and innovation. Standards should cover water efficiency in the production process as well as wastewater and brine management.
- Including water efficiency as a key component of innovation and technology transfer measures relating to renewable hydrogen development. This can include consultations with the mining sector with a view to transferring water stewardship practices from this sector to the renewable hydrogen sector, where applicable.
- Integrating water efficiency as a criterion for assessment and approval of renewable hydrogen projects and for the granting of incentives. In particular, strategic environmental assessments (SEAs) and Environmental Impact Assessments (EIAs) should be used and effectively carried out, the former for assessing renewable hydrogen policies, programmes and plans, and the latter for renewable hydrogen projects.
- Investigating opportunities for co-location and cross-sectoral co-operation between water and renewable hydrogen with a focus on wastewater use, as well as opportunities for the co-location of hydropower infrastructure and renewable hydrogen production. These opportunities should be assessed first through feasibility studies and detailed economic analysis, and then tested via pilot projects.

R6.2: Embed industrial water use policy supporting sustainable access to and use of water into the broader national water management framework

According to the Mongolian law on water, hydrogen project developers are industrial water users. Adopting the nexus approach from the perspective of water sector management means that water authorities should be aware and ready to manage such new water use, including by building the knowledge and capabilities for doing so. It also means that specific tools of the water management framework that play a role in incentivizing water efficiency and sustainable water use can be strengthened. In particular, a sound water allocation regime that establishes conditions for accessing water resources taking into account the economic and social value of water use will be pivotal. To this end, the government of Mongolia can consider the following measures:

- Improve cross-sectoral coordination at Centre of Government level by strengthening the role of the National Water Council and ensuring coherence across water policies. As plans to develop renewable hydrogen production in the country develop, there will be a need to include this sector in the framework of the Council's activities.

- Enhance institutional capabilities, notably of RBOs, through targeted training and capacity-building exercises as well as adequate financial resourcing. This will be pivotal to ensuring that RBOs have the capacity to deal with additional water use at the local level, should renewable hydrogen develop. Specific awareness of hydrogen-related water issues will need to be built.
- Ensure multi-stakeholder dialogue in water matters with proper level of representations across sectors and better transparency and publicity of processes. In the case of hydrogen, this will participate in the better implementation of SEAs and EIAs and can play a great role in preventing conflicts and ensuring local public support for renewable hydrogen project development.
- Improve water data systems to facilitate evidence-based policymaking and enforcement of regulations, as well as water resource mapping including for industrial purposes. Groundwater resources still need to be better assessed in Mongolia, as there remains knowledge gaps as to how much water there is and whether this water is replenishable or not. While progress is being done in this area, accelerating the building of this knowledge will help assess the feasibility of developing renewable hydrogen production at scale. Improving data will be critical for enabling clear and informed dialogues on possible industrial development targets as discussed in recommendation 6.1. It will also be fundamental for monitoring the effects of climate change on the water sector.
- Consider revising the water tariff system in water allocation regimes for further incentivising water efficiency and addressing water pollution. This can be particularly critical in the South Gobi region where water is submitted to great and increasing pressure. Water tariffs have been on an increasing trend in the past years, but there is still room for a differentiated approach across uses that reflect economic and social value of consumer activity and incentivise water efficiency. This logic should apply to renewable hydrogen production in the future, should this activity develop. In this regard, it will be important to implement the recommendations of made by the OECD in the framework of the National Water Policy Dialogue in Mongolia for improving the approach to water tariffs and establishing fully-fledged water allocation regimes in areas where they are needed.
- Integrate future renewable hydrogen sector needs in water infrastructure planning and management. This could entail adopting a logic of "hub" with the co-location of energy, transport and industrial activity such as mining, which is a key market for hydrogen within Mongolia. This can help reduce infrastructure costs thanks to shared use, enable the development of agglomeration economies and lead to circular economy approaches.

Recommendation 7: Ensure hard infrastructure development can support a future low carbon economy

Overview

Whether Mongolia pursues a renewable hydrogen industry or not, significant investment will be necessary in new and upgraded 'hard' infrastructure to underpin sustainable and low carbon economic growth in Mongolia. The challenge – and opportunity – for policymakers in Mongolia is to ensure that decision-making on infrastructure planning and investment can support different directions of sustainable growth at once, and to maximise the co-benefits of new infrastructure development. This is partly a question of planning and strategic decision-making; in such a large and densely populated country, how can Mongolia ensure that any new infrastructure serves the largest number of constituents most effectively?

At the same time, as discussed throughout this report there are a number of infrastructure needs that are of particular importance to any future hydrogen industry, and understanding where those needs are and how this infrastructure can support domestic users will be key to attracting the necessary investment and ensuring that these investments create a positive impact on the local economy. New infrastructure for renewable energy generation and transmission is one obvious example. As discussed in Chapter 3, significant new generation capacity will need to be installed if Mongolia is to produce renewable hydrogen at scale, while new generation and transmission capacity will be necessary in general to meet rising electricity demand and integrate a higher level of variable renewable energy into the national grid. As discussed in Chapter 4, new infrastructure for water use is another example, while upgrading of industrial infrastructure at the firm-level will need to take place in order for the domestic industrial base to absorb low carbon technologies and fuels.

R7.1: Mainstream potential infrastructure requirements for low carbon fuels into industrial policy planning, including for industrial clusters

The government has ambitious industrial policy ambitions, with these continuously developing through the work of the Standing Committee on Industrialisation Policy and other relevant government bodies. Given the future importance of low carbon fuels for Mongolian industry in the future, in both tradeable exported products and non-tradeable industrial products, there is a strong rationale to ensure that policymaking on industrial policy planning explicitly mainstreams the anticipated infrastructure for low carbon fuel use and its transportation. While there are long-standing discussions on the development of steel, oil refining and cement industries – all of which are hard-to-abate – in Mongolia, there are no explicit plans to integrate low carbon considerations into the supportive infrastructure for these industries (e.g., to ensure that new blast furnaces for steel can be switched to hydrogen or co-fired with hydrogen or ammonia). This does not mean that firms are not exploring low carbon and more environmentally sustainable options on their own – there are firms, for example, in both the mining and cement industry that have integrated 'higher' cost technology and infrastructure options into their operations. That firms are undertaking these investments without any significant policy push to do so is an encouraging development, but a more concerted effort through policymaking and strategic planning can ensure that such investments are the norm for new projects rather than the exception.

R7.2: Support the co-location of renewable energy infrastructure with industrial demand

There is a clear rationale for the government to support the co-location of renewable energy infrastructure with sites of industrial demand. One of the main challenges for the Mongolian authorities with the integration of additional variable renewable energy into the grid is the grid's inflexibility. One consequence of this is that it is difficult for energy-intensive industrial firms to decarbonise their activities through electrification, something which has led certain industrial firms to invest in proprietary co-located renewable energy infrastructure; another, more hydrogen-specific consequence is that it is currently not possible to use the national grid to power electrolysis for hydrogen development, which would be the most competitive option over the longer-term. Linked to R7.1, by sending clear signals through strategic planning, investment policy, and the regulatory framework, the government can encourage public and private sector investment in renewable energy infrastructure that is co-located with industrial users. Along with residential use in Ulaanbaatar, industry represents the biggest demand and source of demand growth for power. The national power system is currently struggling to meet this demand, leading to costly electricity imports from Russia. More dedicated co-located energy infrastructure for industry would lower the demand from power-intensive industrial users on the national grid, thereby allowing more variable renewable energy to gradually be integrated into the national power system (both supporting the decarbonisation of the power system and increasing the capacity of the grid to eventually be used to produce renewable hydrogen) and supporting the decarbonisation of industry through direct electrification.

R7.3: Explore a hub and spoke model for water infrastructure

As with other areas of supportive infrastructure for low carbon industrial growth, there is a clear economic and environmental rationale for co-locating new water infrastructure with demand in the industrial and energy sectors. A 'hub and spoke' model of water infrastructure development could reduce the transportation and processing costs of water for hydrogen production, while creating opportunities to derive new value from co- and by-products of industrial processes (for example, using the oxygen produced via electrolysis as an input for water treatment processes), therefore enabling and reaping the benefits of a circular economy.

Recommendation 8: Align the development Mongolia's renewable hydrogen industry with international standards, certification, and regulatory requirements

Overview

Should Mongolia wish to develop renewable hydrogen for export, then developers will have to ensure that the hydrogen developed in the country is compliant with the emerging rules and standards that will govern international hydrogen markets. At present, there is a lack of awareness in Mongolia – both in the public sector as well as in the hard-to-abate industrial sectors in which hydrogen is being both developed and applied – about these rules and standards. There is therefore a risk that the significant investments that are being mobilised for hydrogen research and development will lead to outcomes that are misaligned with these external standards, limiting the competitiveness of a future Mongolian hydrogen industry. At the same time, while there are a limited number of applications of hydrogen in Mongolia at present, the production and transmission of renewable hydrogen as well as its application in new industrial processes will likely require a significant reassessment of the regulatory framework for hydrogen production and handling.

R8.1: Establish a government body to monitor and engage with international certification and standardisation policy development for renewable hydrogen

A new body for hydrogen certification and standardisation could act as a one-stop-shop for information for local producers. There are significant, unresolved international debates on questions of standardisation and certification of renewable hydrogen. The resolution of these debates will have major implications for the methods and value-chain considerations for renewable hydrogen production, transport, storage, and use. Mongolia should ensure that its national standardisation institutions, notably the National Energy Centre which currently is responsible for the implementation of hydrogen-related regulation and its inspection, have a clear mandate to engage in these debates, to disseminate the necessary knowledge domestically, and are equipped to sufficiently monitor and accredit hydrogen industry projects.

R8.2: Undertake a comprehensive, cross-sectoral regulatory review to ensure that the domestic legal framework is coherent with the production, transportation, storage and use of renewable hydrogen

The production of renewable hydrogen in Mongolia, as well as its use in a broader range of industries and sectors, will require significant revision to the regulatory and legal framework that governs hydrogen and other hazardous materials. The challenge facing Mongolia is not dissimilar to that facing policymakers in many OECD countries, and there is significant scope for Mongolia to participate in and learn from international discussions on hydrogen regulation. Given the transversal regulatory nature of hydrogen production and use – in so far as it affects many different sectors and policy areas – the work that would need to be undertaken to update the legal and regulatory framework for hydrogen is likely to be very substantial and require the mobilisation of significant government capacities. Nevertheless, such an exercise is likely to be of fundamental importance for the future application of hydrogen in Mongolia as well as for its production, and as such this exercise should be a major component of any government policy to develop Mongolia's hydrogen industry in the future.

References

ASDIT (2023), *ASDIT*, https://www.asdit.cl/en/inicio-english/. [16]

Diario Fianciero (2023), *Los detalles de la readjudicacion del Instituto de Tecnologias Limpias de Corfo al consorcio Asdit*, https://www.df.cl/df-lab/innovacion-y-startups/los-detalles-de-la-readjudicacion-del-instituto-de-tecnologias-limpias. [15]

Government of Canada (2020), *Canada Hydrogen Strategy*, Government of Canada, https://natural-resources.canada.ca/climate-change-adapting-impacts-and-reducing-emissions/canadas-green-future/the-hydrogen-strategy/23080. [11]

Government of Chile (2021), *Chile's Green Hydrogen Strategy and investment opportunities*, https://energia.gob.cl/sites/default/files/documentos/green_h2_strategy_chile.pdf. [12]

Government of Chile (2021), *Hidrogeno Verde, una oportunidad para la descarbonización de la industria mineria*, https://energia.gob.cl/sites/default/files/documentos/20211019_hidrogeno_y_mineria.pdf. [13]

Government of Morocco (2021), *FEUILLE DE ROUTE HYDROGÈNE VERT*, https://www.mem.gov.ma/Lists/Lst_rapports/Attachments/36/Feuille%20de%20route%20de%20hydrog%C3%A8ne%20vert.pdf. [8]

Government of Namibia (2022), *Namibia Hydrogen Strategy*, Governmen of Namibia, https://www.ensafrica.com/uploads/newsarticles/0_namibia-gh2-strategy-rev2.pdf. [14]

Government of Oman (2022), *Oman's Net Zero National Strategy for an Orderly Transition*, https://www.ea.gov.om/media/aaslyc3l/oman-net-zero-report-2022_screen.pdf. [10]

IEA (2023), *Mongolia*, Greenhouse Gas Emissions from Energy, IEA, Paris, https://www.iea.org/countries/mongolia. [4]

IEA (2023), *Renewable Hydrogen from Oman*, IEA, https://www.iea.org/reports/renewable-hydrogen-from-oman. [9]

IEA (2018), *Mongolia Renewable Energy Feed-in Tariff*, IEA, Paris, https://www.iea.org/policies/6469-mongolia-renewable-energy-feed-in-tariff. [19]

IMF (2021), *Energy Subsidy Template*, IMF, DC, https://www.imf.org/-/media/Files/Topics/Environment/energy-subsidies/fuel-subsidies-template-2022.ashx. [18]

National Statistics Office of Mongolia (2023), *National Accounts*, https://www.1212.mn/en/statistic/statcate/573052/table-view/DT_NSO_0500_001V1. [5]

OEC (2023), *Mongolia*, OEC, https://oec.world/en/profile/country/mng?depthSelector1=HS6Depth&tradeScaleSelector1=tradeScale2. [7]

OECD (2023), *Driving low-carbon innovation for climate neutrality*, OECD, Paris, https://www.oecd.org/sti/ind/policy-brief-driving-low-carbon-innovations-for-climate-neutrality.pdf. [2]

OECD (2022), *Green Hydrogen Opportunities for Emerging and Developing Economies*, OECD, Paris, https://www.oecd-ilibrary.org/docserver/53ad9f22-en.pdf?expires=1696414509&id=id&accname=ocid84004878&checksum=01FB008F220DF39E4DDDC822F7FFEB07. [17]

OECD, World Bank, ESMAP, GIF, Hydrogen Council (2024), *Scaling Hydrogen Financing for Development*, OECD, https://www.oecd-ilibrary.org/docserver/0287b22e-en.pdf?expires=1718379162&id=id&accname=ocid84004878&checksum=6EC64C640B197C21A893CD7B01B93124. [3]

Rodrik, D. (2004), *Industrial Policy for the Twenty-First Century*. [1]

World Bank (2023), *Mongolia Economic Updated April 2023*, World Bank, DC, https://documents1.worldbank.org/curated/en/099413204212338504/pdf/IDU0c9bb5b690643104ccf091520b7f0565236e4.pdf. [6]

Annex A: Case Studies of Hydrogen Pilot Projects in Central Asia

Introduction

This Annex presents six case studies of pilot hydrogen production projects in Central Asia. Of these, four are located in Mongolia, with one in Kazakhstan and Uzbekistan respectively. The information provided in this Annex has been provided by the project developers through a questionnaire, and where possible supplemented by additional publicly available information. The OECD Sustainable Infrastructure Programme in Asia (SIPA) is very grateful to the project developers for their willingness to engage in this exercise, and for their thoroughness and openness with the information provided. The case studies presented in this Annex are for informational purposes only and the information presented here was up to date as of September 2024.

While it is not the purpose of this Annex to be analytical, a number of themes are worth highlighting for context. The first is that although all the pilot projects aim ultimately to export hydrogen or a hydrogen derivative, they are primarily structured in two parts, with the first part of the pilot focused on securing a domestic off-taker for demonstration and feasibility purposes; almost all the projects recognise that securing finance for a competitive export-oriented development strategy will be contingent in the first instance on demonstrating the technical and economic viability of low-carbon technologies in the domestic market.

The second is that there are a number of key technical and financial parameters missing from the case studies, which reflects the fact that while these pilot projects are ambitious, they generally remain in an early stage of development. The third is that, under appropriate policy conditions, all the pilot project developers are confident that they would be able to produce renewable hydrogen in the 2-3 USD/Kg range that the IEA and OECD consider necessary for the fuel to be competitive in hard-to-abate industries. The fourth and last item is that, in the cases of the Mongolian pilot projects, developers are targeting both the same export market and, more challenging, the same domestic industrial market.

Project 1: Sunsteppe Power/Wolfson Energy LLC, OECC

Name of project	
Photograph or depiction of project	*Location of project*
	Umnugobi province, Mongolia
Key technical and financial characteristics of project	
Total foreseen annual production of H2 when operational (by phase if relevant)	The total production is not finalised yet. However, Sunsteppe will tentatively produce 3MW of heat supply by H2
Project status (pre-feasibility, feasibility undertaken, construction, operational)	Pre-feasibility study status
Start date (envisaged if not yet operational)	2026
Abatement cost tonne CO2 equivalent (tCO2-e)	Not available
Electrolyser production capacity	1.5 MW
Lifetime of each electrolyser	Hours/years
Electricity consumption each electrolyser	Not available
Key source(s) of demand	Industry and public agency Mine sites at remote areas
Industrial-site specific/centrally produced and distributed	Mining
Envisaged production cost target	Not available
Envisaged domestic demand	Not available
Total investment needs	Not available
Name(s) of firms involved	SunSteppe Power/ Wolfson Energy LLC, OECC.

Additional sources consulted: (Petro Matad Limited, 2024[1])

Introduction and project overview

SunSteppe Renewable Energy (SRE), a joint venture between Petro Matad Limited and partners, is developing Mongolia's first hydrogen demonstration project in the South Gobi region, Umnugobi province. The project aims for financial close by late 2024, with construction starting in Spring 2025 and commissioning set for Q1 2026. Situated near major coal and copper mining operations, SRE has signed an exclusive Memorandum of Understanding (MOU) to supply green hydrogen to Oyu Tolgoi, one of the world's largest copper and gold deposits, as well as its associated camps.

The project's financing structure will consist of both equity and debt. Wolfson Energy and Petro Matad will provide equity, with the option to engage third-party equity financing under consideration. Debt financing discussions with potential lenders will take place in the future. While financing through technology providers is a possible option, no active discussions are currently underway, though this remains open for exploration as the project advances.

Skills, capacities, and local job creation

The project is expected to create employment opportunities across its phases, each requiring specific skills and experience. During development, energy project development and management expertise are critical. A team of nine has been hired, including Mongolian energy, financial, legal experts, and administrative staff. The construction phase will require an experienced Engineering, Procurement, and Construction (EPC) contractor familiar with hydrogen technology and energy facility construction. Over 100 jobs are expected, including positions for building the hydrogen plant, access road, transmission line, substation, and camp management. In the operational phase, experienced engineers and technicians will be needed for daily operations, with a permanent on-site workforce of 8-10 people managing hydrogen production.

Policy and regulatory considerations

The Mongolian government has demonstrated support for new energy resources through national laws and regulatory policies. While the current policy framework encourages green growth and energy diversification, the lack of specific hydrogen policies and strategies requires the development of a market and off-takers. The Ministry of Energy is currently assessing hydrogen policies and strategies, as awareness of hydrogen's benefits is still emerging. The establishment of the Mongolian Hydrogen Council has played a key role in facilitating hydrogen project development.

The Ministry of Energy has actively promoted hydrogen projects, including engaging with Japanese technology providers during the Prime Minister's visit to Japan in September 2022. In 2022, Wolfson Energy LLA, in collaboration with the Ministry of Energy, also hosted a hydrogen workshop.

This project has received formal support from the Ministry of Energy, including a backing letter and an MOU.

Project perception of risk factors

Mongolia's abundant wind and solar resources reduce feedstock risks for energy projects. The land for the production site has been secured, and wind masts installed, using bankable wind resource data to minimise uncertainties.

Water presents a higher feedstock risk, but SRE's experience has enabled the development of a robust mitigation strategy.

Internationally recognised hydrogen technology has been procured, ensuring reliability despite this being Mongolia's first hydrogen project. Technology assessments by international consultants and reviews by debt financiers further mitigate risks.

While regulatory challenges may arise due to hydrogen's novelty in Mongolia, the government's support for new energy resources, including hydrogen, fosters confidence in an evolving regulatory environment.

Project 2: Euro Khan Sainshand 'Hydrogen Hub' Project

Name of project	
Photograph or depiction of project	*Location of project*
	Sainshand, South Gobi Region
Key technical and financial characteristics of project	
Total foreseen annual production of H2 when operational (by phase if relevant)	Phase one: 1.5Mt H2/year
Project status (pre-feasibility, feasibility undertaken, construction, operational)	Phase two: 10Mt H2/year
Start date (envisaged if not yet operational)	Phase three: 20Mt H2/year
Abatement cost tonne CO2 equivalent (tCO2-e)	Pre-feasibility
Electrolyser production capacity	Not available
Lifetime of each electrolyser	Not available
Electricity consumption each electrolyser	Phase one: 10MW
Key source(s) of demand	Phase two: 100MW + 10MW
Industrial-site specific/centrally produced and distributed	Phase three: 200MW + 10MW
Envisaged production cost target	Not available
Envisaged domestic demand	Not available
Total investment needs	Pilot phase: Mongolian mining sector, primarily foreign operated
Name(s) of firms involved	Later phases: Steel industry in Inner Mongolia, China

Introduction and project overview

Euro Khan, a Mongolian firm with a diverse portfolio in mining, industry, energy, and investment, is leading the exploration of the "Sainshand Green Hydrogen Hub" project in the South Gobi Region of Mongolia. Currently in the pre-feasibility stage, the project benefits from its partners' expertise in renewable energy and hydrogen.

The project consists of two stages. The first stage would involve producing green hydrogen using curtailed renewable energy from the Sainshand Wind Farm, where Euro Khan has a wind farm. A 10MW electrolyser is planned, with a target of producing 1.5Mt of hydrogen annually for domestic mining sector off-takers. The first stage aims to supply green hydrogen to mine operators in Mongolia, with a focus on off-takers like Oyu Tolgoi, a copper and gold mine operated by Rio Tinto that is located close to the production site. Demand is anticipated from applications such as hydrogen power for housing and electricity within mining operations.

Stage two would see production scaled up and exported to industrial users in Inner Mongolia and China, primarily in the steel industry. This phase includes additional renewable and electrolyser capacity, with plans to produce 20Mt of hydrogen annually. The project will use 5-6 years of wind data from the Sainshand Wind Farm to support the selection of the production location.

Should the project be scaled, the developers envisage a final phase in which the project would expand supply chains to serve larger capacity customers, with production scaling to 20,000 TPA of H2, potentially reaching 100,000 TPA. The business case relies on access to cost-effective renewable electricity from Mongolia's solar and wind resources. Japan's projected need to import over 15 mTPA of hydrogen by 2040 and China's growing demand for green hydrogen highlight the market potential.

For the export-oriented scenario, the project envisions additional pipeline infrastructure to connect with the distribution network in Northern China. Economic viability for pipeline investment depends on guaranteed offtake, with a target of 40Mt per year. Rail transport is an interim solution, utilising existing connections to Erenhot, Inner Mongolia, and industrial sites. Ongoing rail developments between Sainshand and Tavan Tolgoi further improve transportation options, with associated costs of approximately USD 0.5/Kg H2 of H2 over a 500 km range, plus wagon acquisition expenses.

Key stakeholders in the project are members of the Mongolian Hydrogen Council, supporting hydrogen development in the country. The project addresses curtailed renewable energy, which results from grid integration constraints, by using excess power from the wind farm. Through an agreement, Euro Khan plans to purchase up to 15,000 MWhrs of curtailed power. The project aims for a production cost of USD 2/Kg H2.

Financing for the pilot phase is being pursued through international institutions like the EIB and EBRD, contingent on securing a domestic off-take contract.

Project framework conditions and feedstock needs

The developers anticipate that water demand will be met by two groundwater aquifers, though their sustainability remains uncertain. The project's water requirements are modest compared to major mining operations, with an estimated 12 litres of ionized water needed per kilogram of H2 and annual consumption projected to produce 10Mt of H2. The project plans to rely on private investment for water infrastructure, avoiding publicly funded sources and collaborating with water authorities during pre-feasibility.

Transporting H2 requires significant infrastructure due to its low energy density. Initially, H2 would be transported using 350 to 700 bar compression. The developers calculate that Sainshand's rail access offers a transport cost of USD 0.50/Kg H2 over 500 km, including wagon costs. Oyu Tolgoi, Mongolia's largest mine, is a potential major consumer of H2 and oxygen, with electrolysis producing 9 kg of O2 for every kg of H2. A new rail link from Sainshand to Tavan Tolgoi expands the distribution of compressed H2 and O2.

While pipelines offer the lowest operating and capital costs, at USD 0.60/Kg H2 for a 300 km pipeline, they are less flexible and require large point-to-point volumes. A 10 NPS pipeline can deliver up to 40,000 tpa of H2 at 70 bar, and pipelines can serve as storage with compression rates of 20 to 70 bar.

Additional renewable energy capacity, including wind and solar, is necessary as production scales. Existing wind infrastructure and planned expansions will support this growth. A collaboration agreement, such as the MOU between Euro Khan and the Sainshand Wind Farm, underscores the project's focus on strategic partnerships. Targeting non-governmental customers helps secure funding, with access to curtailed power enhancing flexibility and aligning with economic viability.

Skills, capacities, and local job creation

Euro Khan plans to leverage its global engineering partnerships, including collaborations with firms like Fichtner & Co, to manage the design, maintenance, and operation of electrolysers and hydrogen handling equipment. With extensive experience in managing wind turbines and large-scale solar installations, Euro Khan brings significant expertise.

Precise job numbers will be determined following the pre-feasibility study, but larger capacities are expected to create at least 100 jobs across various functions. In the initial phase, infrastructure expansion will focus on hydrogen production, storage, and necessary water infrastructure enhancements.

Policy and regulatory considerations

The government has expressed support for green hydrogen projects through a new revival policy within a mid-term program. This initiative focuses on assessing the capacity of renewable energy sources in Umnugobi, exploring technical solutions for substations, transboundary ultra-high voltage power transmission lines linked to the North-East Asian Integrated Energy Network, and green hydrogen production using solar and wind resources.

During the OECD SIPA Roundtable, government representatives acknowledged challenges, noting that the Law of Development policy and planning creates legal barriers to developing a comprehensive national hydrogen strategy. The absence of such a strategy creates uncertainty around the government's support for green hydrogen development, with the current policy seen as too general and ambiguous.

Project perception of risk factors

Securing renewable supply beyond the curtailed energy from the Sainshand Wind Farm presents a moderate challenge, requiring the identification of additional sources and negotiation of agreements for consistent supply. The technologies involved, including electrolysers, wind, and solar, are well-established and advancing.

Attracting quality offtake contracts is crucial for securing funding in the project's early stages. Regulatory uncertainties pose a significant risk due to the lack of prior regulation, requiring careful navigation of legal frameworks. Additionally, occupational Health and Safety (OH&S) risks must be addressed with robust safety measures and protocols as part of regulatory oversight.

Project 3: MAK-UNICO Hydrogen Production and its Derivatives Project

Name of project	
Photograph or depiction of project	*Location of project*
	Demonstration plant will be located in the Eastern Region of Mongolia, in Dornod province.
Key technical and financial characteristics of project	
Total foreseen annual production of H2 when operational (by phase if relevant)	Demonstration plant
Project status (pre-feasibility, feasibility undertaken, construction, operational)	Phase I – Dimethyl ether (DME) – 189,000 metric tonnes/year
Start date (envisaged if not yet operational)	Phase II – H2 – 39,000 metric tonnes/year
Abatement cost tonne CO2 equivalent (tCO2-e)	The demo plant capacity figure indicated above are based on the scoping study. The actual capacity to be optimised and finalised through the pre-feasibility study development based on economic parameters, market size and considered potential technology portfolio.
Electrolyser production capacity	Pre-feasibility study will end in March 2024
Lifetime of each electrolyser	In September 2022, a decree to form a working group from the office of Mongolian President were issued. The working group includes
Electricity consumption each electrolyser	Not available
Key source(s) of demand	(Working group established)
Industrial-site specific/centrally produced and distributed	August 2023 (Pre-feasibility study contract signed)
Envisaged production cost target	The project development concept includes Carbon Capture Storage (CCS-Phase I) and Carbon Capture Utilisation Storage (CCUS – Phase II) technology integration. Emitted CO2 from the process plant is to be stored in nearby geological formation or supplied for Enhance Oil Recovery application, which the emission of CO2 to the atmosphere is completely prevented.
Envisaged domestic demand	Not available
Total investment needs	Not available

Name(s) of firms involved	Not available

Introduction and project overview

The preliminary feasibility study, conducted between Mongolian firm 'Mongolyn Alt (MAK) LLC' and Japanese firm 'Unico International Corporation,' investigates the production of 'blue' hydrogen through coal gasification with carbon capture and storage (CCS). The study envisions a two-stage project, adapting as the hydrogen market develops.

Phase one focuses on producing dimethyl ether (DME) from syngas for domestic heating fuel, with CO2 captured and sequestered in nearby oil exploration sites. DME is a transportable, low-emissions fuel derived from coal. Phase two shifts to hydrogen production from syngas, targeting export markets. This phase integrates hydrogen production into the demonstration plant, with CO2 capture for potential sequestration, enhanced oil recovery, or methanol production.

The project is aligned with the government's ambition to utilise coal for industrialisation and develop a hydrogen industry while addressing Japan's need for low-cost, low-carbon hydrogen imports. The project is closely affiliated with Mongolia's National Industrialisation Committee, established in May 2023, and supported by a dedicated Working Group led by the CEO of MAK.

The project's initial scope focused on evaluating H2 production from bituminous and brown coal to determine the best option for producing coal-based high-value products for domestic and export markets. While many OECD countries prioritise green hydrogen, Japan promotes 'blue' hydrogen from coal using CCS. The export-oriented second phase depends on keeping Mongolian H2 cost-competitive with established trade routes, particularly between Japan and Australia. Cost estimates from a similar project in Japan and Australia suggest H2 production from coal at USD 2.51/Kg by 2030, while the preliminary study projects Mongolian H2 production costs at USD 4.09-4.29/Kg.

Currently, MAK LLC finances working capital and costs, with plans to clarify investment needs and seek financing, including from international finance organisations, after the pre-feasibility study. UNICO presented preliminary investment needs during the 2023 Mongolia-Japan Business Forum, though these are still being optimised.

Project framework conditions and feedstock needs

The project's exact water consumption will be determined during the pre-feasibility study. The demonstration plant site is approximately 10km from the Kherlen River, and MAK considers the water supply sufficient for the pilot phase. Previous internal studies by MAK in 2012 focused on coal liquefaction and provide insights into water supply solutions.

In addition to the Kherlen River, brown coal mine de-watering and hydrological studies indicate an adequate water supply for large-scale industrial operations with minimal environmental impact. The project is exploring the use of water recovered from de-watering brown coal in hydrogen production.

Electricity for the project is expected to come from the Choibalsan combined heat and power plant, with planned expansions. Brown coal from Mongolia's Eastern region will be the primary feedstock, with quantities to be determined in the pre-feasibility study. Gasification of brown coal for DME and hydrogen production uses proven processes, though supplier selection and detailed engineering design are ongoing tasks for later project phases.

Skills, capacities, and local job creation

The demonstration plant will require a skilled workforce in engineering, supply chain management, business administration, and human resources, with the exact number of employees to be determined during the pre-feasibility study. Training for prospective employees will be available at the Choibalsan National Vocational Training Centre.

The project is expected to create both direct and indirect jobs in Dornod province, particularly in Choibalsan.

In the second phase, focused on hydrogen export to East Asian economies, transportation infrastructure development will be critical. The planned expansion of a railway in Eastern Mongolia will enable transport to China via the Jinzhou seaport, facilitating further shipment by sea and supporting regional economic development and connectivity.

Policy and regulatory considerations

The is closely aligned with the National Industrialisation Committee. The project leader advocates for the creation of a national hydrogen strategy to support the growth of domestic hydrogen production and use in Mongolia, reinforcing its commitment to aligning with national industrial goals and promoting sustainable economic development.

Project perception of risk factors

The demonstration plant is located in eastern Mongolia, near a 1.7-billion-tonne brown coal reserve. Global demand for hydrogen, including blue and green varieties, is significant and expected to grow as the hydrogen market develops. However, in Mongolia, the use of CCS/CCUS for Enhanced Oil Recovery (EOR) or geological storage is untested, and regulatory frameworks for these applications are lacking.

Efficient transportation of hydrogen to international markets is critical, requiring careful consideration of logistics, such as railway gauge selection and minimising boil-off gases during transport. Developing additional liquid hydrogen storage and transport technologies will be essential to meet global market demands and ensure project viability.

Project 4: Gobi H2 project

Name of project	
Photograph or depiction of project	*Location of project*
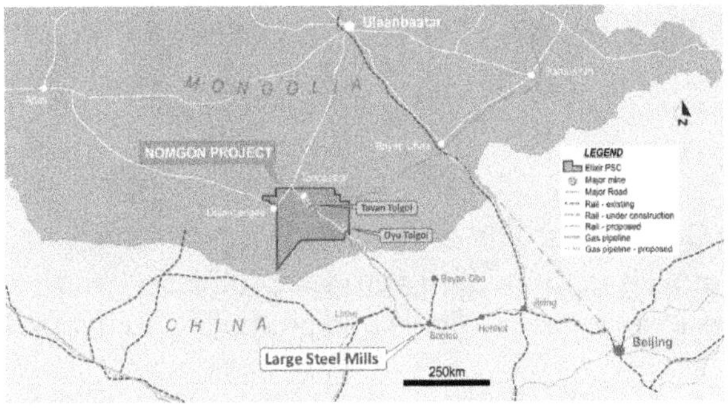	South Gobi region of Mongolia, within the footprint of existing Elixir Energy methane production operations (NOMGON project). The site is proximate to major mine operations in Mongolia (Oyu Tolgoi and Tavan Tolgoi), which are anticipated to be initial off-takers. Located around 400km from Baotou in Inner Mongolia, the project is situated close to the site where China is developing a hydrogen hub, linking the project to potential industrial off-takers in northern China.
Key technical and financial characteristics of project	
Total foreseen annual production of H2 when operational (by phase if relevant)	Not available
Project status (pre-feasibility, feasibility undertaken, construction, operational)	Pre-feasibility study complete, Front-End Engineering Design (FEED) stage commenced. As of August 2023, project contracting engineering consultancy to conduct next stage of feasibility study with an anticipated duration of 12-18months for completion.
Start date (envisaged if not yet operational)	Not defined, but as of August 2023, FID anticipated FEED stage would be finalised in Q4 2024 or Q1 2025. As of September 2024, no publicly available information has been released regarding the start date.
Abatement cost tonne CO2 equivalent (tCO2-e)	Not available
Electrolyser production capacity	Not available
Lifetime of each electrolyser	Not available
Electricity consumption each electrolyser	Not available
Key source(s) of demand	During phase one of operations, focus on mining operators in Mongolia (South Gobi region), primarily multinationals with shareholder and/or regulatory incentives for decarbonisation.

Industrial-site specific/centrally produced and distributed	During phase two of operations, expansion business plan contingent upon offtake agreements with industrial clients in northern China.
Envisaged production cost target	The project estimates green hydrogen production costs at approximately USD 3 per kilogram but has not provided a detailed cost breakdown.
Envisaged domestic demand	Not available
Total investment needs	Not available
Name(s) of firms involved	Stage one: USD 110-130 million

I Additional sources consulted: (AECOM, 2023[2])

Introduction and project overview

The Gobi H2 project is a partnership between Elixir Energy, with over ten years of experience in Mongolia, and Japan's Terras Energy, which is 85% owned by Toyota Tsusho and 15% by SoftBank. The project is located in the South Gobi region, near Elixir Energy's existing methane production area (NOMGON project) and aims to produce green hydrogen using wind and solar energy. It will include renewable energy generation, battery storage, green hydrogen production, and offtake facilities, targeting the rapidly growing Chinese hydrogen market.

The project has completed a pre-feasibility study with international energy consultancy AECOM and is progressing to the Front-End Engineering Design (FEED) stage as of August 2023, pending the securing of engineering consultancy services. Sound detection and ranging (SODAR) towers used during the pre-feasibility study validated the business case and provided critical data for financing negotiations. The project initially targets domestic hydrogen production for the Mongolian mining sector, with plans to export to China for industrial applications in subsequent phases.

Off-take agreements are pending, with ongoing discussions about domestic and Chinese market demands. The project leverages proximity to existing and planned infrastructure, including China's hydrogen hub in Baotou, positioning it strategically for expansion. Challenges remain in water resource management, requiring the development of processing-related infrastructure.

The Gobi H2 project has not disclosed specific demand projections, or the level of Chinese demand needed to proceed with financial decisions or expansion. The project estimates green hydrogen production costs at USD 3 per kilogram but has not provided a detailed cost breakdown.

Domestic demand is expected mainly from Mongolia's mining sector, driven by multinationals' decarbonisation efforts. The project requires pre-agreed off-take contracts before making a final investment decision and has not indicated plans to use Contracts for Difference (CFDs).

The project is engaging with end-users in Mongolia and China and may explore hydrogen derivatives like ammonia, though specifics are not provided. Initial pilot phase investment is estimated at USD 110-130 million, with detailed breakdowns and expansion needs unclear. The project seeks additional external financing, primarily through equity, from sources such as the Mongolian Green Finance Corporation and multilateral banks like the ADB.

Project framework conditions and feedstock needs

The Gobi H2 project is assessing its water requirements, considering insights from nearby mining operations on access, processing, and waste management. While confident in resource adequacy, the project recognises water as a potential risk due to social and political sensitivities in the Gobi region. Plans include co-locating water supply with the hydrogen production facility to streamline operations.

The project will rely on on-site electricity from wind and solar sources. Two scenarios for grid integration are being considered: 1) storing excess renewable energy in batteries or 2) integrating it into the grid. Investment requirements for grid integration depend on capacity upgrades and the development of a new CHP plant.

The technology strategy for the project is in the process of being finalised, upon the receipt of feedback from global industry leaders.

Skills, capacities, and local job creation

In its initial stages, the Gobi H2 project plans to use local transport infrastructure, possibly by road, to deliver hydrogen to mining sector clients. Specific arrangements are pending. For expansion, the project will require new investments in pipeline infrastructure for hydrogen transport between Mongolia and China. The project is consulting multilateral development banks, including the Asian Development Bank (ADB), to support trade connectivity-related infrastructure investments.

Policy and regulatory considerations

The Gobi H2 project has neither sought nor received public funding. However, it is navigating the licensing and permitting process with the support of partners experienced in Mongolia's renewable energy and methane industries. The project submitted a regulatory report on green hydrogen development to Mongolia's Ministry of Energy. Acknowledging the absence of standards for electrolysers and green hydrogen in Mongolia, the project intends to work with the government to create the necessary regulations.

Project perception of risk factors

The Gobi H2 project views the construction of the pilot plant as dependent on securing an offtake contract, primarily targeting mining companies. Securing a guaranteed market for the hydrogen before substantial investment in infrastructure and operations aims to mitigate financial risks.

Project 5: The HYRASIA ONE project

Name of project	
Photograph or depiction of project	*Location of project*
	Mangystau in Western Kazakhstan
Key technical and financial characteristics of project	
Total foreseen annual production of H2 when operational (by phase if relevant)	By 2032, HYRASIA One will produce 2 million tonnes of green hydrogen or 11 million tonnes of green ammonia

Project status (pre-feasibility, feasibility undertaken, construction, operational)	Feasibility
Start date (envisaged if not yet operational)	2032
Abatement cost tonne CO2 equivalent (tCO2-e)	Not available
Electrolyser production capacity	20 GW of electrolyser capacity
Lifetime of each electrolyser	Not available
Electricity consumption each electrolyser	Not available
Key source(s) of demand	Not available
Industrial-site specific/centrally produced and distributed	Not available
Envisaged production cost target	Not available
Envisaged domestic demand	Not available
Total investment needs	With a projected total capital expenditure ranging from USD 40-50 billion, the company plans an initial investment of USD 5-6 billion, with an FID slated for 2026.
Name(s) of firms involved	Svevind Energy group

I Additional sources consulted: (Svevind Energy Group, 2024[3])

Introduction and project overview

The HYRASIA ONE project, led by the German company Svevind, is a large-scale renewable energy initiative in Mangystau, Western Kazakhstan. It plans to establish a 40 GW wind and solar energy complex with 20 GW of electrolyser capacity, producing either two million tonnes of green hydrogen or 11 million tonnes of green ammonia. The project also includes a new desalination plant.

An Investment Agreement was signed with the Kazakh government in October 2022. The pre- Front-End Engineering Design (FEED) stage and Environmental and Social Impact Assessment are scheduled from 2023 to 2026. Pre-FEED work and geological surveys began in June 2023, with construction set for 2027 and full production by 2030. Full operational capacity is expected by 2032.

The HYRASIA ONE project aims to establish Kazakhstan as a key player in European and Asian markets through large-scale green hydrogen production, focusing on both export and domestic use. The project is expected to boost industries like fertiliser manufacturing, chemicals, and cement while promoting regional trade and energy diversification. It will integrate solar and wind energy, hydrogen production via electrolysis, conversion plants, and an export terminal near the port of Kuryk.

With capital expenditures projected at USD 40-50 billion, an initial investment of USD 5-6 billion is planned, with FID expected by 2026. The Investment Agreement with the Kazakh government seeks to attract further investment.

Project framework conditions and feedstock needs

The project's primary feedstocks are solar, wind, and water. Located in an area with high wind speeds and high-quality solar irradiation, detailed resource data has not been disclosed. Water will be sourced and desalinated from the Caspian Sea. A 20 GW electrolyser capacity is planned, supported by a new desalination plant.

Approximately 34,000 square kilometres are reserved for surveys, with 50 square kilometres allocated for the industrial site. The project aims to develop a 40 GW solar and wind farm alongside 20 GW of electrolysis capacity. Information on grid capacity and energy demand remains undisclosed, but electricity generation and hydrogen production will be co-located.

Skills, capacities, and local job creation

The project is expected to create 1,800 permanent jobs and 3,500 during construction. The region's expertise in oil and gas export provides a foundation for transitioning to green hydrogen production. Located near the port of Kuryk on the Caspian Sea, the project will establish a transportation network with a daily capacity of 32,000 metric tonnes, utilising rail, ships, or pipelines. However, limitations to the current transportation network, as outlined below, pose risk to efficiency of export integration.

Policy and regulatory considerations

The Svevind project in Kazakhstan has received strong government support, designated as a landmark initiative for future investments. However, the developers face challenges due to the lack of regulatory frameworks for green hydrogen. Currently, hydrogen is classified solely as an industrial gas under grey chemicals regulation, complicating project approvals. Securing necessary permits is another major hurdle. As the project involves export activities, efforts are underway to meet international regulations, particularly for EU market access. Addressing these regulatory issues is essential for the project's progress.

Project perception of risk factors

Key logistical challenges may pose risks to the project's success. Steep rail gradients in Iran could make transport slower and more expensive, while limited ferry capacity in Turkey may create bottlenecks, delaying or restricting the movement of green ammonia to Europe. These issues could affect the project's efficiency and its ability to meet European export demand.

Project 6: GH2 ACWA Power Project

Name of project	
Photograph or depiction of project	*Location of project*
	Green Hydrogen Plant: located45 km away from Tashkent, in Chirchiq city. Bash 52 MW wind farm: located to the north-eastern part of Ayakagitma reservoir (depression), in Gijduvan district of Bukhara region. Green Ammonia Production: location near either the Navoiy or Sirdarya process plants.

Key technical and financial characteristics of project	
Total foreseen annual production of H2 when operational (by phase if relevant)	3000 tonnes green hydrogen per year
Project status (pre-feasibility, feasibility undertaken, construction, operational)	Not available
Start date (envisaged if not yet operational)	Two Phased COD COD-1 for the HPA by 1st Sep-2024 COD-2 for the PPA by 1st Jan-2025
Abatement cost tonne CO2 equivalent (tCO2-e)	Not available
Electrolyser production capacity	20 MW electrolyser
Lifetime of each electrolyser	Not available
Electricity consumption each electrolyser	Not available
Key source(s) of demand	With National Power Grid of Uzbekistan (NEGU) for 25 years to backstop the purchase of 100% of the renewable energy produced and deliver through wheeling
Industrial-site specific/centrally produced and distributed	Not available
Envisaged production cost target	Not available
Envisaged domestic demand	Not available
Total investment needs	USD ~4 billion
Name(s) of firms involved	ACWA Power – 80%; UKS-20%

Additional sources consulted: (ACWA Power, 2023[4])

Introduction and project overview

In January 2023, Uzbekistan's Ministry of Energy and ACWA Power signed an agreement to develop a green hydrogen facility in Tashkent to reduce reliance on natural gas for fertiliser production. ACWA Power (ACWA), a Saudi-based developer with assets valued at USD 78.2 billion in 77 countries, has been active in Uzbekistan since 2019. The company is now focusing on green hydrogen and ammonia production to support Uzbekistan's industrial decarbonisation and align with the forthcoming Carbon Border Adjustment Mechanism (CBAM).

ACWA Power is developing two projects in Uzbekistan: a green hydrogen production facility and a green ammonia plant.

Green H2 Production: The hydrogen facility will be located 45 km from Tashkent, near the Uzkimyosanoat/MAXAM-Chirchiq-Navoi fertiliser plant. It will use a 20MW electrolyser to produce 3,000 tonnes of green hydrogen annually, with an estimated investment of USD 85-100 million and production expected to begin by late 2024. In addition to the hydrogen plant, ACWA Power will develop a 52 MW wind farm (WF) in Bash, Bukhara region, with a capacity of 80MW to supply power to the grid and support the hydrogen facility. This wind farm will be co-located with the ACWA Power Bash 500 MW WF that is currently under construction. The projects will be managed by a joint venture between ACWA Power and Uzkimyosanoat JSC (UKS), with ACWA holding an 80% stake and UKS holding 20%. UKS will handle licensing, permits, local engineering, land procurement, and secure 10% of the hydrogen offtake through its affiliate while facilitating grid connectivity. The joint venture, ACWA Power UKS Green H2, will oversee development, financing, construction, operation, and maintenance of the wind farm and related infrastructure. A 15-year Hydrogen Purchase Agreement has been signed with Uzkimyosanoat, a state-owned enterprise, and a Power Purchase Agreement has been signed with the National Power Grid of Uzbekistan (NEGU).

Green Ammonia Production: The green ammonia project will be executed in two phases, targeting the production of 500,000 tonnes of green ammonia annually, while reducing gas consumption by 0.6 million MW per year. The facility will be located near either the Navoiy or Sirdarya process plants. The project, estimated to cost USD 3.5-4 billion, includes the construction of 2 GW of renewable energy capacity and 1 GW of electrolysers. A Joint Development Agreement was signed in 2023. The green ammonia project accounts for most of the investment, with USD 1 billion in capital expenditure and USD 1-2 billion in operating expenses over 30 years. The project aims to attract USD 4 billion in foreign direct investment, though detailed cost breakdowns have not yet been disclosed.

Details regarding foreign and domestic demand, as well as Levelised Production Costs, are yet to be disclosed.

Project framework conditions and feedstock needs

For the Green Hydrogen Project, renewable energy and water will be the primary feedstocks. Specific details about water requirements or supply are not yet disclosed. The project intends to expand the existing Bash Wind project, generating 52 MW, which will deliver 175 GWh to the national grid. This electricity will be directed to the Hydrogen Production Unit. The produced green hydrogen will be sold to the Uzkimyosanoat green urea plant

The Green Ammonia Project will require approximately 2 GW of renewable energy sources and 1 GW of electrolysis capacity, which will be developed as part of the project, to produce 500,000MT of green ammonia. It is not specified whether the renewable energy facilities will be co-located with the green ammonia plant or if energy transmission will occur via the national grid, similar to the Green Hydrogen Project. The water requirement for this project is estimated at around 4,500 m³/day, primarily for use in electrolysis.

Skills, capacities, and local job creation

The Green Hydrogen and Green Ammonia projects are expected to create jobs in both construction and operational phases.

The Green Hydrogen Project will generate approximately100 jobs during construction. Up to 50% of the workforce will be recruited locally during peak construction. During the operational phase, approximately 16 new jobs will be created at the wind facility and 10 new jobs will be created for green hydrogen productions. The co-location of the Bash 52 MW WF and the Bash 500 MW WF will enable a total of 35 to 40 workers to be engaged in the operation and maintenance of both wind farms.

The Green Ammonia Project is projected to generate around 3,000 jobs during its four-year construction phase and an estimated 200 to 250 jobs sustained over its ten-year operational phase.

Existing infrastructure will be leveraged and retrofitted to support Green Hydrogen production. Green Ammonia Project may require additional infrastructure, such as grid strengthening, which are included in the project scope. Although transportation infrastructure for fertilisers is already in place, new infrastructure components, including storage, fabrication yards, and other facilities, will be essential.

Policy and regulatory considerations

The establishment of a conducive regulatory environment for the project required collaboration among government stakeholders. Building on the achievements of Phase 1, further developments, such as the establishment of an International Renewable Energy Certificates (iRECs) market, have been initiated. Government support was crucial in overcoming challenges associated with the newness of the industry. The commitment of top leadership has been instrumental in creating an enabling environment for the continuation of these projects.

Project financing, risk allocation, and overall project structuring presented challenges that were addressed with the support of policies from the Ministry of Energy and the Ministry of Industry and Information Technology of Uzbekistan.

Project perception of risk factors

Impact assessments and mitigation approaches have been conducted by ACWA Power for various factors, including terrestrial ecology, air quality, noise and vibration, soils, geology, groundwater and surface water, traffic and transportation, infrastructure and utilities, archaeology and cultural heritage, landscape and visual amenity, shadow flicker, solid waste and wastewater management, socioeconomics, community health and safety, and security, as well as labour and working conditions and influx impact assessment.

Examples of key risks and their corresponding mitigation strategies include:

Habitat Loss: Construction activities such as clearing and excavation pose a risk to vegetation and wildlife, potentially degrading regional ecosystems. To mitigate this, a 30-meter buffer zone will be maintained around wind turbines and substations, and construction footprints will be limited for access roads and substations to minimise disruption.

Air Quality: Dust and pollutant emissions from construction activities can negatively impact both vegetation and wildlife. To address this, a Biodiversity Management Plan has been developed, outlining impact assessments and mitigation strategies to protect the environment and surrounding communities.

These measures are designed to minimise environmental impact and ensure responsible project development.

References

ACWA Power (2023), *Bash 52MW Wind Farm Republic of Uzbekistan: Environmental & Social Impact Assessment Volume 2 - Addendum to the Bash 500MW WF ESIA*. [4]

AECOM (2023), *Gobi Green H2 Prefeasibility Study*, https://aecom.com/au/projects/gobi-green-h2-prefeasibility-study/. [2]

Petro Matad Limited (2024), *Petro Matad Limited Operational Update*, https://markets.ft.com/data/announce/full?dockey=1323-16439481-5TDTQH3AKNPG7UR7F088D307FA. [1]

Svevind Energy Group (2024), *About Hyrasia One*, https://hyrasia.one/the-project. [3]

www.ingramcontent.com/pod-product-compliance
Lightning Source LLC
Chambersburg PA
CBHW061215240926
51846CB00006B/1237